Learning with
Support Vector Machines

Synthesis Lectures on Artificial Intelligence and Machine Learning

Editors
Ronald J. Brachman, *Yahoo! Research*
Thomas Dietterich, *Oregon State University*

Essentials of Game Theory: A Concise Multidisciplinary Introduction
Kevin Leyton-Brown and Yoav Shoham
2008

A Concise Introduction to Multiagent Systems and Distributed Artificial Intelligence
Nikos Vlassis
2007

Intelligent Autonomous Robotics: A Robot Soccer Case Study
Peter Stone
2007

Learning with Support Vector Machines

Colin Campbell and Yiming Ying

ISBN: 978-3-031-00424-7   paperback
ISBN: 978-3-031-01552-6   ebook

DOI 10.1007/978-3-031-01552-6

A Publication in the Springer series
SYNTHESIS LECTURES ON ARTIFICIAL INTELLIGENCE AND MACHINE LEARNING

Lecture #10
Series Editors: Ronald J. Brachman, *Yahoo! Research*
              Thomas Dietterich, *Oregon State University*   Series ISSN
Synthesis Lectures on Artificial Intelligence and Machine Learning
Print 1939-4608   Electronic 1939-4616

Learning with
Support Vector Machines

Colin Campbell
University of Bristol

Yiming Ying
University of Exeter

SYNTHESIS LECTURES ON ARTIFICIAL INTELLIGENCE AND MACHINE LEARNING #10

ABSTRACT

Support Vectors Machines have become a well established tool within machine learning. They work well in practice and have now been used across a wide range of applications from recognizing handwritten digits, to face identification, text categorisation, bioinformatics and database marketing. In this book we give an introductory overview of this subject. We start with a simple Support Vector Machine for performing binary classification before considering multi-class classification and learning in the presence of noise. We show that this framework can be extended to many other scenarios such as prediction with real-valued outputs, novelty detection and the handling of complex output structures such as parse trees. Finally, we give an overview of the main types of kernels which are used in practice and how to learn and make predictions from multiple types of input data.

KEYWORDS

Support Vector Machine, kernels, classification, regression, prediction, learning, generalization, data integration, optimization

Contents

Preface

Support Vectors Machines have become a well established tool within machine learning. They work well in practice and have now been used across a wide range of applications from recognizing hand-written digits, to face identification, text categorisation, bioinformatics and database marketing. Conceptually, they have many advantages justifying their popularity. The whole approach is systematic and properly motivated by statistical learning theory. Training a Support Vector Machine (SVM) involves optimization of a *concave* function: there is a unique solution. This contrasts with various other learning paradigms, such as neural network learning, where the underlying model is generally non-convex and we can potentially arrive at different solutions depending on the starting values for the model parameters. The approach has many other benefits, for example, the constructed model has an explicit dependence on a subset of the datapoints, the *support vectors*, which assists model interpretation. Data is stored in the form of *kernels* which quantify the similarity or dissimilarity of data objects. Kernels can now be constructed for a wide variety of data objects from continuous and discrete input data, through to sequence and graph data. This, and the fact that many types of data can be handled within the same model makes the approach very flexible and powerful. The *kernel substitution* concept is applicable to many other methods for data analysis. Thus SVMs are the most well known of a broad class of methods which use kernels to represent data and can be called *kernel-based methods*.

In this book we give an introductory overview of this subject. We start with a simple Support Vector Machine for performing binary classification before considering multi-class classification and learning in the presence of noise. We show that this framework can be extended to many other scenarios such as prediction with real-valued outputs, novelty detection and the handling of complex output structures such as parse trees. Finally, we give an overview of the main types of kernels which are used in practice and how to learn and make predictions from multiple types of input data.

Colin Campbell and Yiming Ying
February 2011

Acknowledgments

This book originated from an introductory talk on Support Vector Machines given by Colin Campbell (see *www.videolectures.net*). We thank all those who have contributed to our understanding of this subject over the years and to our many collaborators, particularly Nello Cristianini, Tijl de Bie and other members of the Intelligent Systems Laboratory, University of Bristol. A special thanks to Simon Rogers of the University of Glasgow and Mark Girolami, Massimiliano Pontil and other members of the Centre for Computational Statistics and Machine Learning at University College London. We would also like to thank Ding-Xuan Zhou of the City University of Hong Kong.

Colin Campbell and Yiming Ying
February 2011

CHAPTER 1

Support Vector Machines for Classification

1.1 INTRODUCTION

The best way of introducing Support Vector Machines (SVMs) is to consider the simple task of binary classification. Many real-world problems involve prediction over two classes. Thus we may wish to predict whether a currency exchange rate will move up or down, depending on economic data, or whether or not a client should be given a loan based on personal financial information. An SVM is an abstract *learning machine* which will learn from a *training data set* and attempt to *generalize* and make correct predictions on novel data. For the training data we have a set of input vectors, denoted $\mathbf{x}_i$, with each input vector having a number of component *features*. These input vectors are paired with corresponding *labels*, which we denote y_i, and there are m such pairs ($i = 1, \ldots, m$). For our Case Study 2, considered later, we use an SVM to predict disease relapse versus non-relapse for a particular cancer, based on genetic data. In this example, relapse cases would be labeled $y_i = +1$, non-relapse $y_i = -1$, and the matching $\mathbf{x}_i$ are input vectors encoding the genetic data derived from each patient i. Typically, we would be interested in quantifying the performance of the SVM before any practical usage, and so we would evaluate a *test error* based on data from a *test set*.

The training data can be viewed as labelled datapoints in an input space which we depict in Figure 1.1. For two classes of well separated data, the learning task amounts to finding a *directed hyperplane*, that is, an oriented hyperplane such that datapoints on one side will be labelled $y_i = +1$ and those on the other side as $y_i = -1$. The directed hyperplane found by a Support Vector Machine is intuitive: it is that hyperplane which is maximally distant from the two classes of labelled points located on each side. The closest such points on both sides have most influence on the position of this separating hyperplane and are therefore called *support vectors*. The separating hyperplane is given as $\mathbf{w} \cdot \mathbf{x} + b = 0$ (where $\cdot$ denotes the inner or scalar product). b is the *bias* or offset of the hyperplane from the origin in input space, $\mathbf{x}$ are points located within the hyperplane and the normal to the hyperplane, the *weights* $\mathbf{w}$, determine its orientation.

Of course, this picture is too simple for many applications. In Figure 1.1 we show two labelled clusters which are readily *separable* by a hyperplane, which is simply a line in this 2D illustration. In reality, the two clusters could be highly intermeshed with overlapping datapoints: the dataset is then *not linearly separable* (Figure 1.3). This situation is one motivation for introducing the concept of *kernels* later in this chapter. We can also see that stray datapoints could act as anomalous support vectors with a significant impact on the orientation of the hyperplane: we thus need to have a

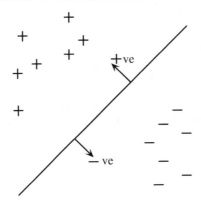

Figure 1.1: The argument inside the decision function of a classifier is $\mathbf{w} \cdot \mathbf{x} + b$. The separating hyperplane corresponding to $\mathbf{w} \cdot \mathbf{x} + b = 0$ is shown as a line in this 2-dimensional plot. This hyperplane separates the two classes of data with points on one side labelled $y_i = +1$ ($\mathbf{w} \cdot \mathbf{x} + b \geq 0$) and points on the other side labelled $y_i = -1$ ($\mathbf{w} \cdot \mathbf{x} + b < 0$).

mechanism for handling noisy and anomalous datapoints. Then again, we need to be able to handle multi-class data.

1.2   SUPPORT VECTOR MACHINES FOR BINARY CLASSIFICATION

Statistical learning theory is a theoretical approach to understanding learning and the ability of learning machines to generalize. From the perspective of statistical learning theory, the motivation for considering binary classifier SVMs comes from a theoretical upper bound on the *generalization error*, that is, the theoretical prediction error when applying the classifier to novel, unseen instances. This generalization error bound has two important features:

[A] the bound is minimized by maximizing the *margin*, γ, i.e., the minimal distance between the hyperplane separating the two classes and the closest datapoints to the hyperplane (see Figure 1.2),

[B] the bound does not depend on the dimensionality of the space.

Let us consider a binary classification task with datapoints $\mathbf{x}_i$ ($i = 1, \ldots, m$), having corresponding labels $y_i = \pm 1$ and let the *decision function* be:

$$f(\mathbf{x}) = \text{sign} (\mathbf{w} \cdot \mathbf{x} + b) \tag{1.1}$$

where $\cdot$ is the scalar or inner product (so $\mathbf{w} \cdot \mathbf{x} \equiv \mathbf{w}^T \mathbf{x}$).

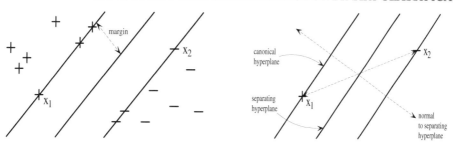

Figure 1.2: *Left*: The perpendicular distance between the separating hyperplane and a hyperplane through the closest points (the support vectors) is called the *margin*, γ. $\mathbf{x}_1$ and $\mathbf{x}_2$ are examples of *support vectors* of opposite sign. The hyperplanes passing through the support vectors are the *canonical hyperplanes*, and the region between the canonical hyperplanes is the *margin band*. *Right*: the projection of the vector $(\mathbf{x}_1 - \mathbf{x}_2)$ onto the normal to the separating hyperplane $(\mathbf{w}/\|\mathbf{w}\|_2)$ is 2γ.

From the decision function we see that the data is correctly classified if $y_i(\mathbf{w} \cdot \mathbf{x}_i + b) > 0 \ \forall i$ since $(\mathbf{w} \cdot \mathbf{x}_i + b)$ should be positive when $y_i = +1$, and it should be negative when $y_i = -1$. The decision function is invariant under a positive rescaling of the argument inside the *sign*-function. This leads to an ambiguity in defining a concept of distance or margin. Hence we implicitly define a scale for $(\mathbf{w}, b)$ by setting $\mathbf{w} \cdot \mathbf{x} + b = 1$ for the closest points on one side and $\mathbf{w} \cdot \mathbf{x} + b = -1$ for the closest on the other side. The hyperplanes passing through $\mathbf{w} \cdot \mathbf{x} + b = 1$ and $\mathbf{w} \cdot \mathbf{x} + b = -1$ are called *canonical hyperplanes*, and the region between these canonical hyperplanes is called the *margin band*. Let $\mathbf{x}_1$ and $\mathbf{x}_2$ be two points inside the canonical hyperplanes on both sides (see Figure 1.2). If $\mathbf{w} \cdot \mathbf{x}_1 + b = 1$ and $\mathbf{w} \cdot \mathbf{x}_2 + b = -1$, we deduce that $\mathbf{w} \cdot (\mathbf{x}_1 - \mathbf{x}_2) = 2$. For the separating hyperplane $\mathbf{w} \cdot \mathbf{x} + b = 0$, the normal vector is $\mathbf{w}/\|\mathbf{w}\|_2$ (where $\|\mathbf{w}\|_2$ is the square root of $\mathbf{w}^T\mathbf{w}$). Thus the distance between the two canonical hyperplanes is equal to the projection of $\mathbf{x}_1 - \mathbf{x}_2$ onto the normal vector $\mathbf{w}/\|\mathbf{w}\|_2$, which gives $(\mathbf{x}_1 - \mathbf{x}_2) \cdot \mathbf{w}/\|\mathbf{w}\|_2 = 2/\|\mathbf{w}\|_2$ (Figure 1.2 (right)). As half the distance between the two canonical hyperplanes, the margin is therefore $\gamma = 1/\|\mathbf{w}\|_2$. Maximizing the margin is therefore equivalent to minimizing:

$$\frac{1}{2}\|\mathbf{w}\|_2^2 \tag{1.2}$$

subject to the constraints:

$$y_i(\mathbf{w} \cdot \mathbf{x}_i + b) \geq 1 \qquad \forall i \tag{1.3}$$

This is a constrained optimization problem in which we minimize an *objective function* (1.2) subject to the *constraints* (1.3). For those unfamiliar with constrained optimization theory we give an outline of relevant material in the Appendix (Chapter A). In the following, we will frequently make use of Lagrange multipliers, Karush-Kuhn-Tucker (KKT) conditions and duality and some familiarity with this subject will be assumed.

As a constrained optimization problem, the above formulation can be reduced to minimization of the following *Lagrange function*, consisting of the sum of the objective function and the m constraints multiplied by their respective *Lagrange multipliers* (see Section A.3). We will call this the *primal* formulation:

$$L(\mathbf{w}, b) = \frac{1}{2}(\mathbf{w} \cdot \mathbf{w}) - \sum_{i=1}^{m} \alpha_i \left(y_i (\mathbf{w} \cdot \mathbf{x}_i + b) - 1\right) \tag{1.4}$$

where α_i are Lagrange multipliers, and thus $\alpha_i \geq 0$ (see Chapter A.3). At the minimum, we can take the derivatives with respect to b and $\mathbf{w}$ and set these to zero:

$$\frac{\partial L}{\partial b} = -\sum_{i=1}^{m} \alpha_i y_i = 0 \tag{1.5}$$

$$\frac{\partial L}{\partial \mathbf{w}} = \mathbf{w} - \sum_{i=1}^{m} \alpha_i y_i \mathbf{x}_i = 0 \tag{1.6}$$

Substituting $\mathbf{w}$ from (1.6) back into $L(\mathbf{w}, b)$, we get the *dual formulation*, also known as the *Wolfe dual*:

$$W(\alpha) = \sum_{i=1}^{m} \alpha_i - \frac{1}{2} \sum_{i,j=1}^{m} \alpha_i \alpha_j y_i y_j \left(\mathbf{x}_i \cdot \mathbf{x}_j\right) \tag{1.7}$$

which must be *maximized* with respect to the α_i subject to the constraints:

$$\alpha_i \geq 0 \qquad \sum_{i=1}^{m} \alpha_i y_i = 0 \tag{1.8}$$

Duality is described in further detail in Section A.2. Whereas the primal formulation involves minimization, the corresponding dual formulation involves maximization and the objective functions of both formulations can be shown to have the same value when the solution of both problems has been found. At the solution, the gradients of the objective function with respect to model parameters (e.g., (1.5) and (1.6)) are zero and hence we use (1.6) to link the two formulations. The dual objective in (1.7) is *quadratic* in the parameters α_i and thus it is described as a *quadratic programming* or *QP* problem. Indeed, given the additional constraints (1.8), it is an example of *constrained quadratic programming*.

So far, we haven't used the second observation, [B], implied by the generalization theorem mentioned above: the generalization error bound does not depend on the dimensionality of the space. From the dual objective (1.7), we notice that the datapoints, $\mathbf{x}_i$, only appear inside an inner product. To get an alternative representation of the data, we could therefore map the datapoints into a space with a different dimensionality, called *feature space*, through a replacement:

$$\mathbf{x}_i \cdot \mathbf{x}_j \rightarrow \Phi(\mathbf{x}_i) \cdot \Phi(\mathbf{x}_j) \tag{1.9}$$

where $\Phi(\cdot)$ is the mapping function. One reason for performing this mapping is that the presented data may not be linearly separable in *input space* (Figure 1.3). It may not be possible to find a directed hyperplane separating the two classes of data, and the above argument fails. Data which is not separable in input space can always be separated in a space of high enough dimensionality. We can visualize this with Figure 1.3 (left): in a 2-dimensional space the two classes cannot be separated by a line. However, with a third dimension such that the $+1$ labelled points are moved forward and the -1 labelled moved back the two classes become separable. Observation [B] means there is no loss of generalization performance if we map to a feature space where the data is separable and a margin can be defined.

The functional form of the mapping $\Phi(\mathbf{x}_i)$ *does not need to be known* since it is implicitly defined by the choice of *kernel*: $K(\mathbf{x}_i, \mathbf{x}_j) = \Phi(\mathbf{x}_i) \cdot \Phi(\mathbf{x}_j)$ or inner product in feature space. Of course, there are restrictions on the possible choice of kernel - a topic we will revisit in Chapter 3. One apparent restriction is that there must be a consistently defined inner product in feature space: this restricts the feature space to being an *inner product space* called a *Hilbert space*.

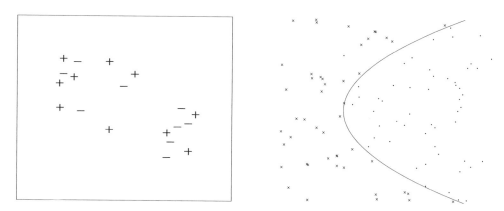

Figure 1.3:   *Left*: in two dimensions these two classes of data are mixed together, and it is not possible to separate them by a line: the data is *not linearly separable*. However, if we imagine a third dimension with the $+1$ labelled points mapped forward from the page and the -1 mapped back, then the two classes become separable by a hyperplane in the plane of the page: with a mapping to this 3D space, the data is now separable. *Right*: using a Gaussian kernel, these two classes of data (labelled ⋆ and ·) become separable by a hyperplane in feature space, which maps to the nonlinear boundary shown, back in input space.

Thus suppose we are presented with a two-class dataset, and we do not know if the data is linearly separable. We could start with a *linear kernel*, $K(\mathbf{x}_i, \mathbf{x}_j) = \mathbf{x}_i \cdot \mathbf{x}_j$, with no mapping to feature space. If the data is not linearly separable then we would not get zero training error if we attempt to solve the optimization problem in (1.7,1.8): some of the training points would be misclassified. We illustrate learning of a non-linearly separable dataset in Case Study 1 (Section 1.6).

Though the data may not be separable in input space, it becomes separable in a higher-dimensional space if we use a *Gaussian kernel* instead (see Section 3.2):

$$K(\mathbf{x}_i, \mathbf{x}_j) = e^{-(\mathbf{x}_i - \mathbf{x}_j)^2 / 2\sigma^2} \tag{1.10}$$

This is illustrated in Figure 1.3 (right). Whereas the two classes of data are not separable with a line in the original 2D space, we are able to separate the two classes using a Gaussian kernel. The two classes of data are separable by a hyperplane in feature space, which maps back to the nonlinear boundary shown in input space. Thus we can achieve zero training error with a Gaussian kernel.

The introduction of a kernel with its implied mapping to feature space is known as *kernel substitution*. Many other choices for a kernel function are possible, e.g.,:

$$K(\mathbf{x}_i, \mathbf{x}_j) = (\mathbf{x}_i \cdot \mathbf{x}_j + 1)^d \qquad\qquad K(\mathbf{x}_i, \mathbf{x}_j) = \tanh(\beta \mathbf{x}_i \cdot \mathbf{x}_j + b) \tag{1.11}$$

which define other types of classifier, in this case a polynomial and a feedforward neural network classifier. Indeed, the class of mathematical objects which can be used as kernels is very general and we consider this topic further in Chapter 3. Furthermore, there is a wide range of methods in machine learning where kernel substitution can also be applied. Many other classification algorithms were previously restricted to linearly separable datasets, but with the use of kernel substitution, they can handle non-separable datasets. Thus the binary-class SVM formulation in (1.7,1.8) is only one instance of a broad range of *kernelizable* methods.

For binary classification with a given choice of kernel, the learning task therefore involves maximization of:

$$W(\alpha) = \sum_{i=1}^{m} \alpha_i - \frac{1}{2} \sum_{i,j=1}^{m} \alpha_i \alpha_j y_i y_j K(\mathbf{x}_i, \mathbf{x}_j) \tag{1.12}$$

subject to the constraints (1.8). The bias, b, has not featured so far, so it must be found separately. For a datapoint with $y_i = +1$, we note that:

$$\min_{\{i | y_i = +1\}} [\mathbf{w} \cdot \mathbf{x}_i + b] = \min_{\{i | y_i = +1\}} \left[\sum_{j=1}^{m} \alpha_j y_j K(\mathbf{x}_i, \mathbf{x}_j) \right] + b = 1 \tag{1.13}$$

using (1.6), with a similar expression for datapoints labeled $y_i = -1$. From this observation we deduce:

$$b = -\frac{1}{2}\left[\max_{\{i|y_i=-1\}}\left(\sum_{j=1}^{m}\alpha_j y_j K(\mathbf{x}_i, \mathbf{x}_j)\right) + \min_{\{i|y_i=+1\}}\left(\sum_{j=1}^{m}\alpha_j y_j K(\mathbf{x}_i, \mathbf{x}_j)\right)\right] \quad (1.14)$$

Thus to construct an SVM binary classifier, we place the data $(\mathbf{x}_i, y_i)$ into (1.12) and maximize $W(\alpha)$ subject to the constraints (1.8). From the optimal values of α_i, which we denote $\alpha_i^\star$, we calculate the bias b from (1.14). For a novel input vector $\mathbf{z}$, the predicted class is then based on the sign of:

$$\phi(\mathbf{z}) = \sum_{i=1}^{m} \alpha_i^\star y_i K(\mathbf{x}_i, \mathbf{z}) + b^\star \quad (1.15)$$

where $b^\star$ denotes the value of the bias at optimality. This expression derives from the substitution of $\mathbf{w}^\star = \sum_{i=1}^{m} \alpha_i^\star y_i \Phi(\mathbf{x}_i)$ (from (1.6)) into the decision function (1.1), i.e., $f(\mathbf{z}) = sign(\mathbf{w}^\star \cdot \Phi(\mathbf{z}) + b^\star) = sign(\sum_i \alpha_i^\star y_i K(\mathbf{x}_i, \mathbf{z}) + b^\star)$. In the following, we will sometimes refer to a solution, $(\alpha^\star, b^\star)$, as a *hypothesis* modelling the data.

When the maximal margin hyperplane is found in feature space, only those points which lie closest to the hyperplane have $\alpha_i^\star > 0$, and these points are the *support vectors*. All other points have $\alpha_i^\star = 0$, and the decision function is independent of these samples. We can see this visually from Figure 1.2: if we remove some of the non-support vectors, the current separating hyperplane would remain unaffected.

From the viewpoint of optimization theory, we can also see this from the *Karush-Kuhn-Tucker* (KKT) conditions: the complete set of conditions which must be satisfied at the optimum of a constrained optimization problem. The KKT conditions are explained in the Appendix (Chapter A.3) and will play an important role in our subsequent discussion of Support Vector Machines. In the absence of a mapping to feature space, one of the KKT conditions (A.28) is:

$$\alpha_i \left(y_i (\mathbf{w} \cdot \mathbf{x}_i + b) - 1\right) = 0 \quad (1.16)$$

from which we deduce that either $y_i (\mathbf{w} \cdot \mathbf{x}_i + b) > 1$ (a non-support vector) and hence $\alpha_i = 0$ or $y_i (\mathbf{w} \cdot \mathbf{x}_i + b) = 1$ (a support vector) allowing for $\alpha_i > 0$.

Datapoints with large values of $\alpha_i^\star$ have a correspondingly large influence on the orientation of the hyperplane and significant influence in the decision function. They could be correct but unusual datapoints, or they could correspond to incorrectly labeled or anomalous examples. When training is complete, it is worth looking at these unusual datapoints to see if they should be discarded: a process called *data cleaning*. Even a correctly labeled outlier has a potential undue influence and so the best approach is to lessen its impact through the use of the *soft margin* approach considered shortly.

Further reading: theoretical generalization bounds for SVMs were first established by Vladimir Vapnik, see, e.g., Vapnik [1998]. Recently, these error bounds were reinterpreted and significantly improved using mathematical tools from approximation theory, functional analysis and statistics, see Cucker and Zhou [2007] and Steinwart and Christmann [2008].

1.3 MULTI-CLASS CLASSIFICATION

Many problems involve multiclass classification and a number of schemes have been outlined. The main strategies are as follows:

- if the number of classes is small then we can use a *directed acyclic graph* (DAG) with the learning task reduced to binary classification at each node. The idea is illustrated in Figure 1.4. Suppose we consider a 3-class classification problem. The first node is a classifier making the binary decision, label 1 versus label 3, say. Depending on the outcome of this decision, the next steps are the decisions 1 versus 2 or 2 versus 3.

- we could use a series of one-against-all classifiers. We construct C separate SVMs with the cth SVM trained using data from class c as the positively labelled samples and the remaining classes as the negatively labelled samples. Associated with the cth SVM we have $f_c(\mathbf{z}) = \sum_i y_i^c \alpha_i^c K(\mathbf{z}, \mathbf{x}_i) + b^c$, and the novel input $\mathbf{z}$ is assigned to the class c such that $f_c(\mathbf{z})$ is largest. Though a popular approach to multi-class SVM training, this method has some drawbacks. For example, suppose there are 100 classes with the same number of samples within each class. The C separate classifiers would each be trained with 99% of the examples in the negatively labelled class and 1% in the positively labelled class: these are very imbalanced datasets, and the multi-class classifier would not work well unless this imbalance is addressed.

- in Section 2.4 we consider *one class classifiers*. Frequently, one class classification is used for novelty detection: the idea is to construct a boundary around the normal data such that a novel point falls outside the boundary and is thus classed as abnormal. The normal data is used to derive an expression ϕ which is positive inside the boundary and negative outside. One class classifiers can be readily adapted to multiclass classification. Thus we can train one-class classifiers for each class c, and the relative ratio of ϕ_c gives the relative confidence that a novel input belongs to a particular class.

Further reading: the DAGSVM method was introduced in Platt et al. [2000]. It is an example of a *one-versus-one* strategy proposed earlier by Hastie and Tibshirani [1998]. An empirical study by Duan and Keerthi [2006] suggested that best performance for multi-class SVM classification was found using the method in Hastie and Tibshirani [1998] together with determination of class membership probabilities using the method discussed in Section 2.3 (see Platt [1999b]). Lee et al. [2001] investigated a variant of *one-versus-all* learning to correct for the imbalance problem mentioned. A method for multi-class classification was proposed by Weston and Watkins [1999] which involved maximization of dual objective functions for multiple separating hyperplanes. An alternative approach based on error-correcting outputs has also been proposed (Allwein et al. [2000], Dietterich and Bakiri [1995]).

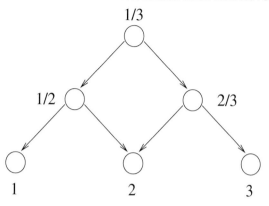

Figure 1.4: With DGSVM a multi-class classification problem is reduced to a series of binary classification tasks.

1.4 LEARNING WITH NOISE: SOFT MARGINS

Most real life datasets contain noise and an SVM can fit to this noise leading to poor generalization. As remarked earlier, outliers can have an undue influence on the position of the separating hyperplane (Figure 1.1). The effects of outliers and noise can be reduced by introducing a *soft margin*, and two schemes are commonly used:

- With an L_1 *error norm*, the learning task is the same as in (1.12, 1.8), except for the introduction of the *box constraint*:

$$0 \leq \alpha_i \leq C \tag{1.17}$$

- With an L_2 *error norm*, the learning task is (1.12, 1.8), except for addition of a small positive constant to the leading diagonal of the kernel matrix

$$K(\mathbf{x}_i, \mathbf{x}_i) \leftarrow K(\mathbf{x}_i, \mathbf{x}_i) + \lambda \tag{1.18}$$

The effect of both these types of soft margin is illustrated in Figure 1.5.

The appropriate values of these parameters can be found by means of a *validation study*. With sufficient data, we would split the dataset into a *training set*, a *validation set* and a *test set*. With regularly spaced values of C or λ, we train the SVM on the training data and find the best choice for this parameter based on the validation error. With more limited data, we may use *cross-validation*, or rotation estimation, in which the data is randomly partitioned into subsets and rotated successively as training and validation data. This would lead to less bias if the validation set is small.

The justification for using soft margin techniques comes from statistical learning theory and can be viewed as relaxation of the *hard margin* constraint (1.3). In effect, we allow for some

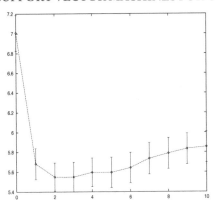 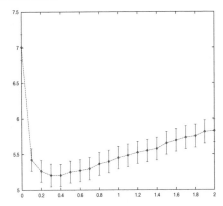

Figure 1.5: Soft margin classification using L_1 (left) and L_2 (right) error norms. *Left*: test error as a percentage (y-axis) versus C (x-axis). *Right*: test error as a percentage (y-axis) versus λ (x-axis). The ionosphere dataset from the UCI Machine Learning Repository [UCI Machine learning Repository] was used with Gaussian kernels ($\sigma = 1.5$).

datapoints inside the margin band or even on the wrong side of the hyperplane during training: the latter possibility means we have a non-zero training error. Datapoints inside the margin band or on the wrong side of the hyperplane are called *margin errors*.

L_1 **error norm**: For the L_1 *error norm*, we introduce a non-negative *slack variable* ξ_i into (1.3):

$$y_i \left(\mathbf{w} \cdot \mathbf{x}_i + b \right) \geq 1 - \xi_i \tag{1.19}$$

and the task is now to minimize the sum of errors $\sum_{i=1}^{m} \xi_i$ in addition to $||\mathbf{w}||^2$:

$$\min \left[\frac{1}{2} \mathbf{w} \cdot \mathbf{w} + C \sum_{i=1}^{m} \xi_i \right] \tag{1.20}$$

If $\xi_i > 0$, we have a margin error. The above is readily formulated as a primal Lagrange function:

$$
\begin{aligned}
L(\mathbf{w}, b, \alpha, \xi) \;=\; & \frac{1}{2} \mathbf{w} \cdot \mathbf{w} + C \sum_{i=1}^{m} \xi_i \\
& - \sum_{i=1}^{m} \alpha_i \left[y_i \left(\mathbf{w} \cdot \mathbf{x}_i + b \right) - 1 + \xi_i \right] - \sum_{i=1}^{m} r_i \xi_i
\end{aligned}
\tag{1.21}
$$

with Lagrange multipliers $\alpha_i \geq 0$ to handle (1.19) and $r_i \geq 0$ to handle the requirement $\xi_i \geq 0$. At the optimum, the derivatives with respect to $\mathbf{w}$, b and ξ give:

$$\frac{\partial L}{\partial \mathbf{w}} = \mathbf{w} - \sum_{i=1}^{m} \alpha_i y_i \mathbf{x}_i = 0 \tag{1.22}$$

$$\frac{\partial L}{\partial b} = \sum_{i=1}^{m} \alpha_i y_i = 0 \tag{1.23}$$

$$\frac{\partial L}{\partial \xi_i} = C - \alpha_i - r_i = 0 \tag{1.24}$$

Using the first of these to eliminate $\mathbf{w}$ from $L(\mathbf{w}, b, \alpha, \xi)$, we obtain the same dual objective function, (1.12), as before. However, $r_i \geq 0$ and $C - \alpha_i - r_i = 0$, hence $\alpha_i \leq C$, and the constraint $0 \leq \alpha_i$ is replaced by $0 \leq \alpha_i \leq C$. Patterns with values $0 < \alpha_i < C$ will be referred to as *non-bound*, and those with $\alpha_i = 0$ or $\alpha_i = C$ will be said to be *at-bound*. For an L_1 error norm, we find the bias in the decision function (1.15) from the KKT conditions, which include the following two equations

$$r_i \xi_i = 0 \tag{1.25}$$

$$\alpha_i (y_i (\mathbf{w} \cdot \mathbf{x}_i + b) - 1 + \xi_i) = 0 \tag{1.26}$$

Suppose we select a *non-bound* datapoint k, so $0 < \alpha_k < C$. From $C - \alpha_k - r_k = 0$ and $\alpha_k < C$, we deduce that $r_k > 0$ and hence that $\xi_k = 0$ from (1.25). Since $\alpha_k > 0$, we deduce from (1.26) that $y_k(\mathbf{w} \cdot \mathbf{x}_k + b) = 1$. Finally, with $y_k^2 = 1$ and $\mathbf{w} = \sum_{i=1}^{m} \alpha_i y_i \mathbf{x}_i$, we see that the bias can be obtained by averaging:

$$b = y_k - \sum_{i=1}^{m} \alpha_i y_i (\mathbf{x}_i \cdot \mathbf{x}_k) \tag{1.27}$$

over all such non-bound examples.

L_2 **error norm**: For the L_2 *error norm* the objective function is:

$$\min \left[\frac{1}{2} \mathbf{w} \cdot \mathbf{w} + C \sum_{i=1}^{m} \xi_i^2 \right] \tag{1.28}$$

subject to:

$$y_i (\mathbf{w} \cdot \mathbf{x}_i + b) \geq 1 - \xi_i, \qquad \xi_i \geq 0. \tag{1.29}$$

Introducing Lagrange multipliers to incorporate the constraints, we get a primal Lagrange function:

$$L(\mathbf{w}, b, \alpha, \xi) = \frac{1}{2}\mathbf{w} \cdot \mathbf{w} + C \sum_{i=1}^{m} \xi_i^2$$

$$-\sum_{i=1}^{m} \alpha_i \left[y_i \left(\mathbf{w} \cdot \mathbf{x}_i + b\right) - 1 + \xi_i\right] - \sum_{i=1}^{m} r_i \xi_i \qquad (1.30)$$

with $\alpha_i \geq 0$ and $r_i \geq 0$. After obtaining the derivatives with respect to $\mathbf{w}$, b and ξ, substituting for $\mathbf{w}$ and ξ in the primal objective function and noting that the dual objective function is maximal when $r_i = 0$, we obtain the following dual objective function after kernel substitution:

$$W(\alpha) = \sum_{i=1}^{m} \alpha_i - \frac{1}{2} \sum_{i,j=1}^{m} y_i y_j \alpha_i \alpha_j K(\mathbf{x}_i, \mathbf{x}_j) - \frac{1}{4C} \sum_{i=1}^{m} \alpha_i^2 \qquad (1.31)$$

With $\lambda = 1/2C$ this gives the same dual objective function as for hard margin learning except for the substitution (1.18).

Asymmetric soft margins: For many real-life datasets there is an imbalance between the amount of data in different classes, or the significance of the data in the two classes can be quite different. For example, for the detection of tumors on magnetic resonance imaging (MRI) scans, it may be best to allow a higher number of false positives if this improved the true positive detection rate. The relative balance between the detection rate for different classes can be easily shifted by introducing *asymmetric soft margin parameters*. Thus for binary classification with an L_1 error norm, $0 \leq \alpha_i \leq C_+$ ($y_i = +1$), and $0 \leq \alpha_i \leq C_-$ ($y_i = -1$), while $K(\mathbf{x}_i, \mathbf{x}_i) \leftarrow K(\mathbf{x}_i, \mathbf{x}_i) + \lambda_+$ (if $y_i = +1$) and $K(\mathbf{x}_i, \mathbf{x}_i) \leftarrow K(\mathbf{x}_i, \mathbf{x}_i) + \lambda_-$ (if $y_i = -1$) for the L_2 error norm.

ν-**SVM**: A problem with the L_1 and L_2-norm soft margin SVMs is that there is no clear interpretation associated with the parameters C and λ. With an alternative approach, called ν-SVM, we use a soft margin parameter ν, which can be related to the fraction of datapoints which have margin errors and the fraction which are support vectors. νSVM is a variant on the L_1 *error norm* soft margin classifier with the following primal formulation:

$$\min_{\mathbf{w}, \xi, \rho} \left\{ L(\mathbf{w}, \xi, \rho) = \frac{1}{2}\mathbf{w} \cdot \mathbf{w} - \nu\rho + \frac{1}{m} \sum_{i=1}^{m} \xi_i \right\} \qquad (1.32)$$

subject to:

$$y_i \left(\mathbf{w} \cdot \mathbf{x}_i + b\right) \geq \rho - \xi_i \qquad (1.33)$$
$$\xi_i \geq 0 \qquad (1.34)$$
$$\rho \geq 0 \qquad (1.35)$$

Introducing Lagrange multipliers α_i, β_i and δ for the three constraint conditions (1.33 - 1.35), we have the primal Lagrange function:

$$L(\mathbf{w}, \xi, b, \rho, \alpha, \beta, \delta) = \frac{1}{2}\mathbf{w} \cdot \mathbf{w} - \nu\rho + \frac{1}{m}\sum_{i=1}^{m}\xi_i \tag{1.36}$$

$$-\sum_{i=1}^{m}(\alpha_i[y_i(\mathbf{w} \cdot \mathbf{x}_i + b) - \rho + \xi_i] + \beta_i\xi_i) - \delta\rho \tag{1.37}$$

where $\alpha_i \geq 0$, $\beta_i \geq 0$ and $\delta \geq 0$. At the optimum, we take the derivatives of L with respect to the four variables $\mathbf{w}$, ξ, b and ρ and set these to zero, giving the following:

$$\mathbf{w} = \sum_{i=1}^{m}\alpha_i y_i \mathbf{x}_i \tag{1.38}$$

$$\alpha_i + \beta_i = \frac{1}{m} \tag{1.39}$$

$$\sum_{i=1}^{m}\alpha_i y_i = 0 \tag{1.40}$$

$$\sum_{i=1}^{m}\alpha_i - \delta = \nu \tag{1.41}$$

In addition, we have the KKT conditions, which give the following conditions at the solution:

$$\beta_i\xi_i = 0 \tag{1.42}$$

$$\delta\rho = 0 \tag{1.43}$$

$$\alpha_i[y_i(\mathbf{w} \cdot \mathbf{x}_i + b) - \rho + \xi_i] = 0 \tag{1.44}$$

With $\rho > 0$ condition, (1.43) implies $\delta = 0$. In that case, (1.41) implies $\sum_{i=1}^{m}\alpha_i = \nu$. If we have a margin error ($\xi_i > 0$), then (1.42) implies $\beta_i = 0$. In that case, (1.39) implies $\alpha_i = 1/m$. From $\sum_{i=1}^{m}\alpha_i = \nu$ and $\alpha_i = 1/m$, for a margin error, we deduce that the fraction of margin errors is upper bounded by ν. In addition, since α_i can be at most equal to $1/m$, from (1.39), there must be at least νm support vectors. Of course, this argument is based on the assumption $\rho > 0$. However, we can relate ρ to the non-zero margin between points without a margin error ($\xi_i = 0$). If $\xi_i = 0$, then (1.33) implies $y_i(\mathbf{w} \cdot \mathbf{x}_i + b) \geq \rho$. Thus retracing our argument in Section 1.2, we find that this margin is $\gamma = \rho/\|\mathbf{w}\|_2$, so $\rho > 0$ for a typical learning problem.

In summary, if we use νSVM with $\rho > 0$, then ν has the following interpretation:

[1] ν is an upper bound on the fraction of margin errors ($\xi_i > 0$),

[2] ν is a lower bound on the fraction of support vectors.

To use this method, we need to derive the dual formulation. We eliminate $\mathbf{w}, \xi, b$ and ρ from (1.36) through re-substitutions using (1.38) and (1.39). This gives the dual formulation:

$$\max_{\alpha} \left\{ W(\alpha) = -\frac{1}{2} \sum_{i,j=1}^{m} \alpha_i \alpha_j y_i y_j K(\mathbf{x}_i, \mathbf{x}_j) \right\} \qquad (1.45)$$

subject to:

$$0 \le \alpha_i \quad \le \quad \frac{1}{m} \qquad (1.46)$$

$$\sum_{i=1}^{m} \alpha_i y_i \quad = \quad 0 \qquad (1.47)$$

$$\sum_{i=1}^{m} \alpha_i \quad \ge \quad \nu \qquad (1.48)$$

Thus, for a practical application, we first select ν and solve the constrained quadratic programming problem (1.45 - 1.48) for the given dataset. If $(\alpha_i^\star, b^\star)$ is the solution, then the decision function for a novel datapoint $\mathbf{z}$ is based on the sign of:

$$\phi(\mathbf{z}) = \sum_{i=1}^{m} y_i \alpha_i^\star K(\mathbf{x}_i, \mathbf{z}) + b^\star \qquad (1.49)$$

as before. To find the bias $b^\star$, we find a datapoint, which we label k, with $y_k = +1$ and such that $1/m > \alpha_k > 0$ so that $\xi_k = 0$. From the KKT condition (1.44), we deduce that $(\mathbf{w} \cdot \mathbf{x}_k + b) - \rho = 0$. Using a similar non-bound point, which we label l with $y_l = -1$, we find that:

$$b^\star = -\frac{1}{2} \left[\sum_{j=1}^{m} \alpha_j y_j K(x_j, x_k) + \sum_{j=1}^{m} \alpha_j y_j K(x_j, x_l) \right] \qquad (1.50)$$

Further reading: soft margins were introduced by Cortes and Vapnik [1995] and the ν-SVM by Schölkopf et al. [2000]. Asymmetric soft margins were introduced in Veropoulos et al. [1999].

1.5   ALGORITHMIC IMPLEMENTATION OF SUPPORT VECTOR MACHINES

Quadratic programming (QP) is a standard problem in optimization theory, so there are a number of software resources available. *MATLAB* has the routine *QUADPROG* within its optimization toolbox. Generic quadratic programming packages include *MINOS* and *LOQO*. In addition, there are a number of packages specifically written for SVMs such as *SVMlight*, *LIBSVM* and *SimpleSVM*. For methods we introduce later, such as the linear programming (LP) approaches to classification and novelty detection, there are also many packages available, such as *CPLEX*. Generic QP and LP packages can be used for moderate size problems, but they have the disadvantage that the kernel matrix is stored in memory. For larger datasets, therefore, we must use dedicated techniques which only use a part of the kernel at each step and which are optimized for SVM learning.

Working set methods avoid problems with large-size kernel matrices by using an evolving subset of the data. For example, with *chunking*, we only use a variable-size subset or *chunk* of data at each stage. Thus a QP routine is used to optimize the dual objective on an initial arbitrary subset of data. The support vectors are retained and all other datapoints (with $\alpha_i = 0$) discarded. A new working set of data is then derived from these support vectors and additional unused datapoints. This *chunking* process is then iterated until the margin is maximized. Of course, this procedure will still fail if the dataset is too large. *Decomposition methods* provide a better approach: with these methods we only use a subset of the data of *fixed size* at each step of the method. The α_i of the remaining datapoints are kept fixed.

Further reading: working set and large-scale learning methods for SVMs are reviewed in Bottou et al. [2007]. Software is available from the following sources:

LIBSVM: http://www.csie.ntu.edu.tw/~cjlin/libsvm/

SVMlight: http://svmlight.joachims.org

simpleSVM: http://asi.insa-rouen.fr/~gloosli/simpleSVM.html

SVQP: http://leon.bottou.org/projects/svqp

MINOS: http://www.sbsi-sol-optimize.com/asp/sol_products_minos.htm

LOQO: http://www.princeton.edu/~rvdb/loqo

CPLEX: http://www-01.ibm.com/software/integration/optimization/cplex-optimizer/

Decomposition and Sequential Minimal optimization (SMO): the limiting case of decomposition is the *Sequential Minimal Optimization* (SMO) algorithm in which *only two* α variables are optimized at each step. Since this working set method can efficiently handle large scale problems, we will outline the approach here. Let us suppose that the *pair* of α parameters we will alter are indexed i and j. Furthermore, let indices k and l range over *all* the samples in the following, including i and j. During updating of the pair (α_i, α_j), let us suppose the corresponding corrections are:

$$\begin{aligned} \alpha_i &\rightarrow \alpha_i + \delta_i \\ \alpha_j &\rightarrow \alpha_j + \delta_j \end{aligned}$$

If ΔW is the corresponding change in $W(\alpha)$ and letting K_{kl} denote $K(\mathbf{x}_k, \mathbf{x}_l)$, then:

$$\Delta W = \delta_i + \delta_j - y_i \delta_i \left(\sum_{k=1}^{m} y_k \alpha_k K_{ik} \right) - y_j \delta_j \left(\sum_{k=1}^{m} y_k \alpha_k K_{jk} \right)$$
$$-\frac{1}{2} y_i^2 \delta_i^2 K_{ii} - y_i y_j \delta_i \delta_j K_{ij} - \frac{1}{2} y_j^2 \delta_j^2 K_{jj}$$

In addition, the constraint $\sum_{k=1}^{m} \alpha_k y_k = 0$ must be satisfied throughout, which imposes the condition:

$$\delta_i y_i + \delta_j y_j = 0 \tag{1.51}$$

The working set, the minimal set of variables to alter, cannot be less than two because of this equality constraint. However, we can use the latter constraint to eliminate either δ_i or δ_j and thus solve an optimization problem *in one variable* at each step. Specifically if we eliminate δ_i, calculate the optimum of ΔW with respect to δ_j then, for binary classification (so $y_j^2 = y_i^2 = 1$), we get:

$$\delta_j^{new} = \frac{y_j \left(\sum_{k=1}^{m} y_k \alpha_k K_{ik} - y_i - \sum_{k=1}^{m} y_k \alpha_k K_{jk} + y_j \right)}{K_{ii} + K_{jj} - 2K_{ij}} \tag{1.52}$$

If we define the derivative of the dual objective in (1.12) as:

$$g_l = \frac{\partial W}{\partial \alpha_l} = 1 - y_l \sum_{k=1}^{m} y_k \alpha_k K_{lk} \tag{1.53}$$

and let $\eta = K_{ii} + K_{jj} - 2K_{ij}$, then we can write:

$$\delta_j^{new} = y_j \eta^{-1} \left(y_j g_j - y_i g_i \right) \tag{1.54}$$

with $j \leftrightarrow i$ for δ_i^{new}. The gain in W is given by:

$$\Delta W = \frac{1}{2} \eta^{-1} \left(y_i g_i - y_j g_j \right)^2 \tag{1.55}$$

In the rare instance that $\eta = 0$, we conclude that we cannot make progress with this pair.

If we are using a hard margin or using a soft margin via the L_2 error norm, then we could maximize ΔW by selecting that i which maximizes $y_i g_i$ and that j which minimizes $y_j g_j$. For a soft margin using the L_1 error norm, the box constraint $0 \le \alpha_i \le C$ must be satisfied after selecting the pair and updating α_i and α_j. The box constraint can be written:

$$S_i \le y_i \alpha_i \le T_i \tag{1.56}$$

where $[S_i, T_i] = [0, C]$ if $y_i = 1$ and $[S_i, T_i] = [-C, 0]$ if $y_i = -1$. Let us define index sets $I_{upper} = \{i \mid y_i \alpha_i < T_i\}$ and $I_{lower} = \{j \mid y_j \alpha_j > S_j\}$. From (1.54), a suitable choice is then the pair:

$$i \quad = \quad \max_{k \in I_{upper}} \{y_k g_k\} \tag{1.57}$$

$$j \quad = \quad \min_{k \in I_{lower}} \{y_k g_k\} \tag{1.58}$$

Other choices are possible. For example, from (1.55), we could use a maximal gain criterion:

$$i \quad = \quad \max_{k \in I_{upper}} \{y_k g_k\} \tag{1.59}$$

$$j \quad = \quad \max_{k \in I_{lower}} \frac{1}{2}\eta^{-1}(y_i g_i - y_k g_k)^2 \qquad \text{subject to} \qquad y_i g_i > y_k g_k \tag{1.60}$$

After updating, the α-updates may need to be clipped to satisfy $0 \le \alpha_i \le C$. As an algorithm, we therefore proceed through the dataset looking for the next pair (i, j) to update using (1.57, 1.58) or (1.59, 1.60), for example. This pair of α are updated as in (1.54). This procedure is followed until a stopping criterion, such as:

$$\max_{i \in I_{upper}} \{y_i g_i\} - \min_{j \in I_{lower}} \{y_j g_j\} < \epsilon \tag{1.61}$$

is observed (ϵ is a small tolerance). This solves the optimisation problem in (1.12, 1.8), following which we derive the bias b.

Further reading: SMO was introduced by Platt [1999a] with improvements made by Keerthi et al. [2001] and Shevade et al. [2000]. This method is described further in Bottou et al. [2007] and Schölkopf and Smola [2002a].

1.6   CASE STUDY 1: TRAINING A SUPPORT VECTOR MACHINE

With our first Case Study, we consider the training of a Support Vector Machine for a straightforward binary classification task. Our dataset will be drawn from the *UCI Machine Learning Repository*, which is a useful resource for testing and benchmarking algorithms. In this example, we make use of their *ionosphere dataset*. This dataset has 351 samples, which are classed into *good* (labelled +1) or *bad* (−1) radar returns, each sample having 34 features. We split the data and used the first $m = 300$ datapoints as a training set (written $\mathbf{x}_i$ with corresponding labels $y_i = \pm 1$ where $i = 1, \ldots, 300$), with the remaining 51 datapoints as the test set.

First, we assumed a linear kernel and then attempted to solve the constrained quadratic programming problem in (1.7, 1.8) using a *QP* package, called *CFSQP*, written in C. If we can find a tentative solution for the α_i, we next compute the bias b from equation (1.14). It is always good practice to verify that the program has found a solution by evaluating the training error at the end of the training process; that is, we compute the fraction of training points such that

$y_i(\sum_{j=1}^{m} \alpha_j y_j K(\mathbf{x}_i, \mathbf{x}_j) + b) < 0$ (with a soft margin a few training points could potentially be in error). For hard margin learning, we can also check the margin at the solution since we should find $|\phi(\mathbf{x}_i)| \geq 1$ for all training points $\mathbf{z} = \mathbf{x}_i$ in (1.15). We can also check the expression $\sum_i \alpha_i y_i = 0$ at the proposed solution to verify (1.8). An optional further measure is to evaluate W from (1.12) after a fixed number of iterations of the QP solver. When converging on a reasonable solution, W typically shows a monotonic increase in value with each iteration, apart from some initial fluctuations. The difference in W between iterations, ΔW, typically decreases steadily with each iteration as it approaches the solution. With the choice of a linear kernel, and a hard margin, we found that W showed extended erratic and non-monotonic behaviour with the ionosphere dataset, before apparently converging on a solution with a large value of W. This tentative solution did not have zero training error, nor was the margin 1. This dataset is therefore not linearly separable and we must proceed to a nonlinear kernel.

We next used the Gaussian kernel given in (1.10) which is suitable for even the most complex non-linearly separable dataset. One immediate observation is that poorly chosen values for the kernel parameter σ could lead to spurious performance. Thus if σ is too small, the argument inside the exponential would have a large negative value. On the leading diagonal of $K(\mathbf{x}_i, \mathbf{x}_j)$ there would be values of 1, but the off-diagonal components of the kernel matrix would be effectively 0 (since $\exp(-\infty) \to 0$), and the kernel matrix has too little discriminative leverage to gain a good solution. If σ is too big, then all entries in the kernel matrix become nearly 1, and we similarly loose discriminative power. Writing the Gaussian kernel as $K(\mathbf{x}_i, \mathbf{x}_j) = \exp(-arg_{ij})$, a good choice of σ would give $arg_{ij} \approx O(1)$ for typical training points. This can be found by calculating the average of $(\mathbf{x}_i - \mathbf{x}_j)^2$ and choosing σ accordingly, or by fixing $\sigma = 1$ and normalising the data before use, so that this condition applies. The most appropriate strategy for finding σ, or any other kernel parameter, is to use a validation dataset. We discuss this and other strategies for finding kernel parameters in Section 3.3. Of course, *we must only use the validation data and never the test dataset as a means for finding kernel or soft margin parameters*. Inappropriate use of the test data corrupts the test statistic and can lead to an unfair statement of the test error.

Returning to the ionosphere dataset, with a choice of $\sigma = 1$ for the kernel parameter, and unadjusted data, we achieved zero training error and, on presenting the test data, the SVM gave a test percentage error of 5.9%. The ionosphere dataset does not have missing values. If some values are missing from a dataset, then we need to *impute* these: with *MATLAB* one such procedure is *kNNimpute*.

1.7   CASE STUDY 2: PREDICTING DISEASE PROGRESSION

We now illustrate the use of Support Vector Machines with an application to predicting disease progression. In this study, the objective is to predict relapse versus non-relapse for Wilm's tumor, a cancer which accounts for about 6% of all childhood malignancies. This tumor originates in the kidney but it is a curable disease in the majority of affected children with more than 85% long-term survival after diagnosis. However, there is a recognised aggressive subtype with a high probability of relapse within a few years. It is, therefore, clinically important to predict risk of relapse when the disease is first diagnosed with a more aggressive treatment regime if risk of relapse is high. In this study we used cDNA microarray data as input to the Support Vector Machine. The cDNA microarray had 30,720 probes each measuring *gene expression*, roughly speaking, the protein production rate of individual genes. Many of the readings from the microarray were missing or poor quality so an initial quality filtering reduced the number of *features* to 17,790 reliable readings. The dataset was approximately balanced with 27 samples of which 13 samples were from patients who relapsed with the disease within 2 years, and the remainder labelled as non-relapse due to long-term survival without disease recurrence.

This looks like an obvious instance in which we could use a Support Vector Machine. The task is binary classification, and the large number of features means the relatively small number of datapoints are embedded in a high-dimensional space. A high dimensional space means the separating hyperplane has a large degree of freedom in terms of its orientation and position and thus such datasets are typically linearly separable. Thus we only need use a *linear kernel* $K(\mathbf{x}_i, \mathbf{x}_j) = \mathbf{x}_i \cdot \mathbf{x}_j$. Yet another advantage of using an SVM is that the learning task only involves the number of samples ($m = 27$) and not the number of features. Thus the SVM can be trained rapidly. This contrasts with other approaches to machine learning, such as those based on neural networks, where the inputs are all the measured features per sample and thus training times could become prohibitive.

However, there are evident problems. Firstly, we do not have the luxury of large training and test sets. In instances like this, we must revert to *leave-one-out* (LOO) testing. This is, we train on 26 samples and evaluate classifier performance on a single held-out datapoint. We successively rotate the test datapoint through the data so that each sample is held-out once. This procedure is not without its disadvantages. It is computationally intensive with m classifiers trained in all, and there is a high commonality among the training set per classifier leading to a possible bias. However, with such a small dataset, it is the only route available. The other problem is that the number of features is very large. The vast majority of these features are likely to be irrelevant. Indeed, if we have a few informative features and a large background which is uninformative, then it is likely that the latter will drown the signal from the former and our classifier will exhibit poor generalization. In addition, we would like to remove uninformative features because this will improve model interpretation. By doing so, we highlight those genes which are most relevant to distinguishing relapse from non-relapse.

This application therefore motivates the use of *feature selection* methods, and we will discuss two general approaches. Firstly, *filter methods* in which features are scored individually, using statistical methods, prior to training the classifier. Secondly, we consider a *wrapper method* in which the algorithm uses an internal procedure to eliminate redundant features.

Filter Methods: The choice of filter method can amount to a prior assumption about the way in which we consider a feature as significant. Roughly speaking, filter methods can be viewed as falling into two groupings. Some methods are more influenced by the consistency of a difference between classes and some are more influenced by the magnitude of a difference. For example, a set of 4 samples with one feature each, $(1.1, 1.1, 1.1, 1.1)$, is consistently different from a second set of samples, $(0.9, 0.9, 0.9, 0.9)$, and both can be separated by a simple threshold (1.0). On the other hand, there is a significant difference in the means of the two sets $(1.0, 1.0, 5.0, 5.0)$ and $(1.0, 1.0, 0.1, 0.1)$ even though the first two members of each set are the same.

One type of filter method uses statistical scores based on ranking datapoints. These scores can be more influenced by the consistency of a difference since if all values belonging to one class are ranked higher than all members of the other class; this is determined as an improbable event even if the means of both classes do not differ a great deal. For example, for binary classification, the Mann-Whitney U test provides such a rank-based measure between two populations of datapoints. An alternative is the *Fisher score*, which is more influenced by the magnitude of a difference. A good discriminating feature would have a large separation between means and small standard deviations for datapoints belonging to the two classes. If the means are μ_+ and μ_- and standard deviations σ_+ and σ_-, the Fisher score is $F = (\mu_+ - \mu_-)^2/(\sigma_+^2 + \sigma_-^2)$. For the Wilm's tumor dataset, it is worth considering both types of filter. Abnormal genes can be distinguished by very distinct over- or under-expression, and thus they would be highlighted by a Fisher score. On the other hand, abnormally functioning genes can also have an expression profile typified by a broad spread of expression values. With a standard deviation in the denominator, the Fisher score is at a disadvantage, but these types of profiles could still be highlighted by a Mann-Whitney score. Finally, a balanced and effective score is the t-test for the difference between two populations in which a probability measure can be obtained from Student's distribution.

Recursive Elimination of Features: Rather than a prior scoring of features, we could use a wrapper method to eliminate features during training. For a linear kernel, we note that the weight matrix for an SVM can be expressed as $\mathbf{w} = \sum_{i=1}^{m} y_i \alpha_i \mathbf{x}_i$ from (1.6). The smallest component values of the weight matrix will have least influence in the decision function and will therefore be the best candidates for removal. With the *Recursive Feature Elimination* (RFE) method, the SVM is trained with the current set of features, and the best candidate feature for removal is identified via the weight vector. This feature is removed and the process repeated until termination. One disadvantage of this method is that the process will be slow if there are a large number of features: in this case, features are removed in batches. The algorithm is terminated by using held-out validation data with the best feature set retained and subsequently used when this validation error passes through a minimum.

Application to the Wilm's Tumor Dataset: We now apply these methods to the Wilm's tumor dataset. In Figure 1.6, we show the LOO test error (y-axis) versus number of top-ranked features (x-axis) remaining, using filter methods based on the Fisher score (left) and t-test scoring (right) of features. Since the training set is different for each rotation of the test point with LOO testing, the set of top-ranked features can be different with each such training partition.

For the Fisher score filter the minimal LOO test error is 5/27, for Mann-Whitney, it is 4/27 and for the t-test it is 1/27. RFE gave similar results though partly dependent on whether we eliminated features in batches and the size of these batches. All these approaches indicate prediction is very poor if all features are used. However, good prediction can apparently be achieved with a small number of

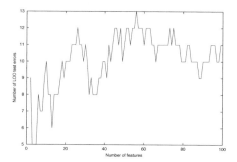
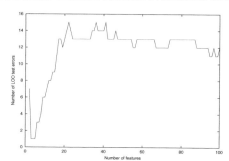

Figure 1.6: The number of LOO test errors (y-axis) versus number of top-ranked features (x-axis) remaining, using a Fisher score filter (*left*) or t-test filter (*right*) for predicting relapse or non-relapse for Wilm's tumor.

features. In all cases, the LOO test error starts rising when a very small number of features are used because the data is no longer separable in this very low dimensional space. The t-test performed best here and in most similar applications we have considered. These results suggest prediction of relapse can be achieved. However, given the small size of the dataset and the substantial noise inherent in expression array data, this result should be treated with caution. Nonetheless, a detailed subsequent investigation indicated the related technology of arrayCGH (comparative genomic hybridisation) could give robust prediction beyond 90% and that relapse was largely conditional on abnormal copy number variation of genetic sequence on several chromosomes.

Further reading: RFE was introduced by Guyon et al. [2002] with a review of feature selection methods in Guyon and Elisseeff [2003]. The Wilm's tumor example is drawn from Williams et al. [2004].

1.8   CASE STUDY 3: DRUG DISCOVERY THROUGH ACTIVE LEARNING

In this second case study, we consider the use of SVMs for the discovery of active compounds which bind to a particular target molecule. This is an important part of the drug discovery process. An iterative procedure is used in which a batch of unlabeled compounds is screened against the given target, and we wish to find as many active *hits* as possible. At each iteration, we would select a set of unlabeled compounds and subsequently label them as positive (active) or negative (inactive).

Just as the last study motivated an interest in feature selection, this problem will motivate a discussion of *passive versus active learning*. With *passive learning*, the learning machine receives examples and learns from these: this has been the situation considered so far. However, with *active learning* the learning machine will pose queries. These should be maximally informative so as to understand the underlying rule generating the data, using the fewest possible examples. The idea behind active learning is appealing: from previously learnt examples we can form a rough hypothesis. This hypothesis will give some guidance as to the next most informative queries to pose. The answers to these will, in turn, improve our hypothesis and therefore the next generation of queries. This is in contrast to passive learning: as randomly generated examples are received, they may contain little new information, or they could repeat previous examples, thus providing no new information at all.

An SVM looks like a suitable engine for active learning since the non-support vectors do not make a contribution to the decision function for classification. Thus if there are few support vectors and we can find all of these early in the training process, we would not need to ask for the labels of the non-support vectors.

Several strategies are possible for active learning. We could create queries. However, sometimes created queries are meaningless and impossible to label (examples readily appear if we create handwritten characters, for example). Here we will only consider *selective sampling* in which we can query the label of currently unlabelled datapoints. For a Support Vector Machine, querying a point within the margin band *always* guarantees a gain in the dual objective $W(\alpha)$ in (1.12), whatever the label. Querying points outside the band only gives a gain in $W(\alpha)$ if the current hypothesis, or decision function, predicts this label incorrectly. Indeed, within the margin band, *the best points to query are those unlabeled datapoints closest to the current hyperplane*. Intuitively, these points are maximally ambiguous and they are likely to become support vectors. We can formally establish this observation as follows. For a hard margin Support Vector Machine, querying a point $\mathbf{x}_k$ adds its index k to the current index set, which we write as $I_{new} = I_{old} \cup \{k\}$. When we have learnt with the current index set, the following KKT condition applies at optimality $\alpha_i^\star (y_i [\mathbf{w} \cdot \mathbf{x}_i + b] - 1) = 0$. From (1.4) in Chapter 1, this gives:

$$W(\alpha^\star) = L(\mathbf{w}^\star) = \frac{1}{2}\mathbf{w}^\star \cdot \mathbf{w}^\star = \frac{1}{2\gamma^2} \tag{1.62}$$

since, at an optimum, the values of the dual and primal objective functions are equal, $W(\alpha^\star) = L(\mathbf{w}^\star)$ (see Chapter A.2). Thus:

$$\Delta W = \frac{1}{2}\left(\frac{1}{\gamma_{I_{new}}^2} - \frac{1}{\gamma_{I_{old}}^2}\right) \tag{1.63}$$

Thus points closest to the current hyperplane ($\gamma_{I_{new}}$ small) are the best choice for increasing the optimized $W(\alpha)$. Starting from a pair of oppositely labeled datapoints, we therefore proceed through the dataset by requesting the label of the next unlabeled datapoint closest to the current separating hyperplane. Of course, we also need a suitable stopping criterion for active learning; otherwise, we learn all the data with no practical gain over passive learning. For noiseless data, we could stop when the dual objective $W(\alpha)$ stops increasing, or we could monitor the cumulative number of prediction mistakes made by the current hypothesis (an example is given in Figure 1.7): that is, the prediction error made on the next unlabelled datapoint as they are successively learnt.

For more typical datasets containing noise, an effective criterion is to stop when the margin band is emptied of unlabeled datapoints (see Figure 1.8 for two examples).

In Figure 1.7 we compare passive learning versus active learning for a simple noise-free example called the *majority rule*. We randomly create strings of 20 binary digits, and the label is +1 if there are more 1's than 0's and −1 otherwise. After selecting about 60 examples (Figure 1.7 (left)), the active learner has learnt the majority rule perfectly with no prediction errors (Figure 1.7 (right)).

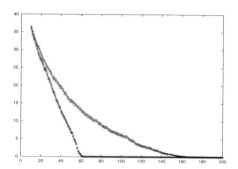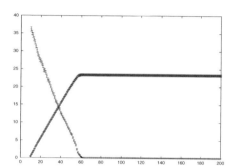

Figure 1.7:   *Left*: test error (y-axis) as a percentage versus number of patterns learnt (x-axis) for passive learning (top curve) versus active learning (bottom curve). *Right*: the curve with diamond centerpoints gives the cumulative number of prediction errors (y-axis) (errors on the predicted label of the next point to query) versus number of patterns learnt (x-axis), the curve with bar centerpoints gives the corresponding test error. This experiment was for the *majority rule* with 20 bits strings and 200 training and 200 test points averaged over 100 experiments.

In Figure 1.8, we compare active versus passive learning for two real-life datasets which contain noise. For active learning, the test error passes through a minimum which co-incides closely with the point at which the current margin band becomes empty: the closest unlabelled datapoint to the current separating hyperplane now lies outside this band. Before this minimum it is learning representative datapoints, which are likely to become support vectors. Beyond the minimum, the learning machine is either learning nothing new (a non-support vector), or it is probably learning deleterious outliers which can have a significant negative impact on the orientation of the separating hyperplane.

Indeed, even when training a Support Vector Machine without consideration of active learning, it is also often beneficial to start the learning process with a few reliable datapoints from each class and then successively learn that datapoint which is closest to the current hyperplane, discarding (not learning) the residual set of datapoints when the current margin band has emptied. This is an alternative to using a soft margin if outliers are believed present, and the gains can be similar to those presented in Figure 1.8.

Finally, we come to our application to drug discovery (Figure 1.9). In this study, each compound was described by a vector of 139,351 binary shape features. Two datasets were considered. For the

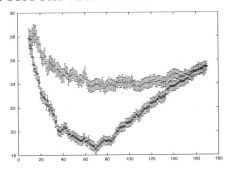
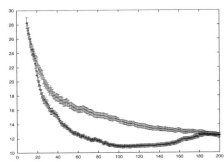

Figure 1.8: *Left*: test error (*y*-axis) as a percentage versus number of patterns learnt (*x*-axis) for passive learning (top curve) versus active learning (bottom curve) for the Cleveland heart dataset from the UCI Database Repository. After an average 70 examples, the margin band is empty, which looks an effective stopping point. *Right*: test error (*y*-axis) as a percentage versus number of patterns learnt (*x*-axis) for passive learning (top curve) versus active learning (bottom curve) for the ionosphere dataset from the UCI Repository. After an average 94 examples, the margin band is empty.

first dataset, there were 39 active compounds among 1,316 chemically diverse examples, and for the second, there were 150 active compounds among 634. The results are portrayed in Figure 1.9.

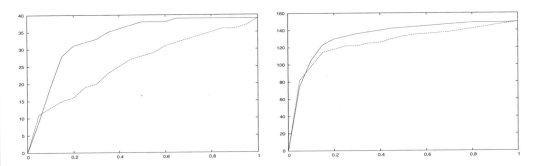

Figure 1.9: The number of active compounds found (*y*-axis) versus fraction of examples selected. The solid curves are for active learning and the dashed curves are passive learning. *Left*: this dataset has 39 active compounds among 1,316 chemically diverse examples. *Right*: this dataset has 150 active compounds among 634 diverse examples.

For both datasets, there is a gain for active learning over passive learning. This gain tends to increase as the size of a dataset increases since the *sparsity ratio* (the number of support vectors over number of datapoints) typically falls as the dataset size increases. Indeed, we notice that the largest gain for active learning is for the first dataset, which is the largest. Of course, the gain made by active learning is at the cost of a computationally time consuming search for the next best candidate compound. For this reason, active learning methods from machine learning have only had a limited impact in drug discovery so far. Nevertheless, the idea of automating the discovery process is conceptually appealing, and we can expect further applications in future.

Further reading: the active learning approach described here was introduced by Campbell et al. [2000] and Tong and Koller [2001]. The application to drug discovery was described in Warmuth et al. [2003].

CHAPTER 2

Kernel-based Models

2.1 INTRODUCTION

In the last chapter, we introduced the idea of kernel substitution for the handling of non-separable datasets. In fact, kernel substitution can be applied to a wide range of data analysis methods so that SVMs should really be viewed as a sub-instance of a much broader class of *kernel-based methods*. After introducing a *linear programming* approach to SVM learning, we extend the capabilities of SVMs in the remainder of the chapter. One current disadvantage of SVMs is that there is no confidence measure associated with the predicted class label. In real-life applications this could be a drawback. For example, in Case Study 2, we used an SVM to predict relapse versus non-relapse for Wilm's cancer. A clinician would plan differently with a reliable high confidence prediction as against a low confidence one. Thus we discuss a simple strategy for introducing a probability measure. Next, we introduce *one class* classification. In this case, the method creates a boundary around the datapoints. One application is to novelty detection in which we label a new datapoint as novel or as belonging to the established data distribution. So far, we have only considered discrete outputs, so we next consider *regression* in which the output is a real-valued number. Finally, we show that the SVM framework is capable of handling much more complex output structures. With *structured output learning*, we consider outputs consisting of ordered data structures such as parse trees.

2.2 OTHER KERNEL-BASED LEARNING MACHINES

Rather than using quadratic programming, it is also possible to derive a kernel classifier in which the learning task involves *linear programming (LP)* instead. For binary classification, the predicted class label is determined by the sign of the following:

$$\phi(\mathbf{z}) = \sum_{i=1}^{m} w_i K(\mathbf{z}, \mathbf{x}_i) + b \tag{2.1}$$

In contrast to an SVM where we used a 2-norm for the weights (the sum of the squares of the weights), here we will use a 1-*norm* (the sum of the absolute values):

$$||\mathbf{w}||_1 = \sum_{i=1}^{m} |w_i| \tag{2.2}$$

This norm is useful since it encourages sparse solutions in which many or most weights are zero. During training, we minimize an objective function:

$$L = \frac{1}{m} ||\mathbf{w}||_1 + C \sum_{i=1}^{m} \xi_i \tag{2.3}$$

where the slack variable ξ_i is defined by:

$$\xi_i = \max\{1 - y_i\phi(\mathbf{x}_i), 0\} \tag{2.4}$$

These slack variables are, therefore, positively-valued when $y_i\phi(\mathbf{x}_i) < 1$, which occurs when there is a *margin error*. Furthermore, we can always write any variable w_i, potentially positive or negative, as the difference of two positively-valued variable $w_i = \alpha_i - \widehat{\alpha}_i$ where $\alpha_i, \widehat{\alpha}_i \geq 0$. To make the value of w_i as small as possible, we therefore minimize the sum $(\alpha_i + \widehat{\alpha}_i)$. We thus obtain a linear programming (LP) problem with objective function:

$$\min_{\alpha,\xi,b} \left[\frac{1}{m} \sum_{i=1}^{m} (\alpha_i + \widehat{\alpha}_i) + C \sum_{i=1}^{m} \xi_i \right] \tag{2.5}$$

with $\alpha_i, \widehat{\alpha}_i, \xi_i \geq 0$ and subject to constraints (from 2.4):

$$y_i\phi(\mathbf{x}_i) \geq 1 - \xi_i \tag{2.6}$$

The predicted label is therefore based on the sign of:

$$\phi(\mathbf{x}_i) = \sum_{j=1}^{m} \left(\alpha_j^\star - \widehat{\alpha}_j^\star \right) K(\mathbf{x}_i, \mathbf{x}_j) + b^\star \tag{2.7}$$

where $(\alpha_j^\star, \widehat{\alpha}_j^\star, b^\star)$ are the values of these parameters at the optimum.

Further reading: linear programming approaches to learning have been considered by a number of authors (Bennett [1999], Bradley and Mangasarian [2000], Graepel et al. [1999], Mangasarian [1965]). There are other approaches to kernel-based learning which can offer possible advantages. In the Appendix (Section A.1), we introduce the concept of *version space*: the geometric dual of input space. From the viewpoint of version space, the solution found by an SVM is not optimal, and a better solution is derived from using the *center of mass* and *Bayes point* in version space. Kernel-based methods which follow this approach are the *Analytic Center Machine* (Malyscheff and Trafalis [2002], Trafalis and Malyscheff [2002]) and the *Bayes Point Machine* (Harrington et al. [2003], Herbrich et al. [2001], Minka [2001], Rujan and Marchand [2000]). Kernel substitution can be applied to many other techniques such as the Fisher discriminant (Baudat and Anouar [2000], Mika et al. [1999], Roth and Steinhage [2000], Schölkopf and Smola [2002a]) and least squares approaches (Suykens and Vandewalle [1999]). All methods discussed so far are for *supervised learning*: the learning machine is trained from examples with known labels. Of course, we may also be interested in the *unsupervised* discovery of structure within data. Examples of unsupervised methods where kernel substitution can be used include *kernel PCA* (principal component analysis) (Schölkopf and Smola [2002a], Shawe-Taylor and Cristianini [2004]) and *kernel CCA* (canonical correlation analysis) (Shawe-Taylor and Cristianini [2004]).

2.3   INTRODUCING A CONFIDENCE MEASURE

If used for predicting diagnostic categories, for example, it would be useful to have a confidence measure for the class assignment in addition to determining the class label. For binary classification, an SVM does have an inbuilt measure that could be exploited to provide a confidence measure for the class assignment, i.e., the distance of a new point from the separating hyperplane (Figure 1.2). A test point with a large distance from the separating hyperplane should be assigned a higher degree of confidence than a point which lies close to the hyperplane.

Recall that the output of a SVM, before thresholding to ± 1, is given by

$$\phi(\mathbf{z}) = \sum_i y_i \alpha_i K(\mathbf{x}_i, \mathbf{z}) + b \tag{2.8}$$

One approach is to fit the *posterior probability* $p(y|\phi)$ directly. A good choice for mapping function is the *sigmoid*:

$$p(y = +1|\phi) = \frac{1}{1 + \exp(A\phi + B)} \tag{2.9}$$

with the parameters A and B found from the training set $(y_i, \phi(\mathbf{x}_i))$. Let us define t_i as the target probabilities:

$$t_i = \frac{y_i + 1}{2} \tag{2.10}$$

so for $y_i \in \{-1, 1\}$ we have $t_i \in \{0, 1\}$. We find A and B by performing the following minimization over the entire training set:

$$min_{A,B} \left[-\sum_i t_i \log(p_i) + (1 - t_i) \log(1 - p_i) \right] \tag{2.11}$$

where p_i is simply (2.9) evaluated at $\phi(\mathbf{x}_i)$. This is a straightforward 2-dimensional nonlinear minimization problem, which can be solved using a number of optimization routines. Once the sigmoid has been found using the training set, we can use (2.9) to calculate the probability that a new test point belongs to either class. Figure 2.1 shows the training values and fitted sigmoid from a cDNA microarray dataset for ovarian cancer. Note that no datapoints are present in a band between $+1$ and -1, due to the use of a hard margin and the data being linearly separable.

Further reading: The above method was proposed by Platt [1999b]. To develop a fully probabilistic approach to kernel-based learning requires some understanding of Bayesian methods, which takes

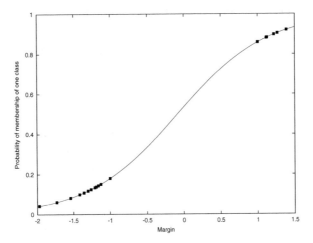

Figure 2.1: Probability of membership of one class (y-axis) versus margin. The plot shows the training points and fitted sigmoid for an ovarian cancer data set. A hard margin was used, which explains the absence of points in the central band between -1 and $+1$.

us beyond this introductory text. *Gaussian processes* (GPs) are a kernel-based approach, which give a probability distribution for prediction on novel test points. Ignoring the bias term b, the real-valued output of the GP is $y_i = \mathbf{w}^T \Phi(\mathbf{x}_i)$ where $\mathbf{w}$ is the weight vector. We follow a probabilistic approach with an assumed prior probability distribution over the weights $\mathbf{w}$, namely that they are approximately normally distributed and thus modeled by a Gaussian distribution $p(\mathbf{w}) = N(\mathbf{w}|\mathbf{0}, \beta^{-1}\mathbf{I})$ where β is the *precision* (inverse variance) of the distribution. From $\mathbf{y} = \Phi\mathbf{w}$ and the assumed distribution for $p(\mathbf{w})$, we infer that $\mathbf{y}$ follows a Gaussian distribution itself. The mean of the probability distribution for $\mathbf{y}$ is given by $E[\mathbf{y}] = \Phi E[\mathbf{w}] = 0$ and the covariance matrix is $E[\mathbf{y}\mathbf{y}^T] = \frac{1}{\beta}\Phi\Phi^T = \mathbf{K}$ where we define the *kernel function*:

$$K_{ij} = \frac{1}{\beta}\Phi(\mathbf{x}_i)^T \Phi(\mathbf{x}_j)$$

With a novel input $\mathbf{z}$ this approach gives a mean value for the prediction and an associated spread. A full description of Gaussian Processes is presented in Rasmussen and Williams [2006] and Schölkopf and Smola [2002a].

2.4   ONE CLASS CLASSIFICATION

For many real-world problems, the task is not to classify but to detect novel or abnormal instances. Novelty detection has potential applications in many application domains such as condition monitoring or medical diagnosis. For example, suppose we are looking for tumors in magnetic resonance imaging (MRI) scans. If we are classifying image objects, there is a high risk of failure in certain cases. Thus tumors can have unusual shapes and a new instance may be very unlike any object in the training set. A novelty detector would act as a secondary screen since an unusual shape may not be classified correctly, but a novelty detector should still highlight the shape as *abnormal*.

One approach to novelty detection is *one-class classification* in which the task is to model the *support* of a data distribution, i.e., to create a function which is positive in those regions of input space where the data predominantly lies and negative elsewhere. Aside from an application to novelty detection, one class classification has other potential uses. As mentioned earlier, it can be used for multi-class classification. Suppose, for class $c = 1$, we create a function $\phi_{c=1}$ which is positive where class $c = 1$ data is located and negative elsewhere, then we could create a set of ϕ_c for each data class and the relative ratios of the ϕ_c would decide for ambiguous datapoints not falling readily into one class.

One approach to one class learning is to find a hypersphere with a minimal radius R and center $\mathbf{a}$ which contains all or most of the data: novel test points lie outside the boundary of this hypersphere. The effect of outliers is reduced by using slack variables ξ_i to allow for outliers outside the sphere, and the task is to minimize the volume of the sphere and number of datapoints outside i.e.,

$$\min \left[R^2 + \frac{1}{m\nu} \sum_i \xi_i \right] \tag{2.12}$$

subject to the constraints:

$$(\mathbf{x}_i - \mathbf{a})^T (\mathbf{x}_i - \mathbf{a}) \leq R^2 + \xi_i \tag{2.13}$$

and $\xi_i \geq 0$, and where ν controls the tradeoff between the two terms. The primal formulation is then:

$$
\begin{aligned}
L(R, \mathbf{a}, \alpha_i, \xi_i) \;=\; & R^2 + \frac{1}{m\nu} \sum_{i=1}^{m} \xi_i - \sum_{i=1}^{m} \gamma_i \xi_i \\
& - \sum_{i=1}^{m} \alpha_i \left(R^2 + \xi_i - (\mathbf{x}_i \cdot \mathbf{x}_i - 2\mathbf{a} \cdot \mathbf{x}_i + \mathbf{a} \cdot \mathbf{a}) \right)
\end{aligned} \tag{2.14}
$$

with $\alpha_i \geq 0$ and $\gamma_i \geq 0$. After kernel substitution the dual formulation gives rise to a *quadratic programming* problem, namely maximize:

$$W(\alpha) = \sum_{i=1}^{m} \alpha_i K(\mathbf{x}_i, \mathbf{x}_i) - \sum_{i,j=1}^{m} \alpha_i \alpha_j K(\mathbf{x}_i, \mathbf{x}_j) \tag{2.15}$$

with respect to α_i and subject to $\sum_{i=1}^{m} \alpha_i = 1$ and $0 \leq \alpha_i \leq 1/mv$. If $mv > 1$ then *at bound* examples will occur with $\alpha_i = 1/mv$ and these correspond to outliers in the training process. Having completed the training process, a test point $\mathbf{z}$ is declared *novel* if:

$$\phi(\mathbf{z}) = R^2 - K(\mathbf{z}, \mathbf{z}) + 2 \sum_{i=1}^{m} \alpha_i K(\mathbf{z}, \mathbf{x}_i) - \sum_{i,j=1}^{m} \alpha_i \alpha_j K(\mathbf{x}_i, \mathbf{x}_j) < 0 \qquad (2.16)$$

R^2 is determined by finding a datapoint k which is *non-bound*, i.e., $0 < \alpha_k < 1/mv$ and setting this inequality to an equality, i.e., $\phi(\mathbf{z} = \mathbf{x}_k) = 0$.

   An alternative approach uses *linear programming*. As above, the objective is to find a surface in input space which wraps around the data clusters: anything outside this surface is viewed as abnormal. This surface can be defined as the *level set*, $\phi(\mathbf{z}) = 0$, of a nonlinear function. We map the data to feature space $\mathbf{x_i} \rightarrow \Phi(\mathbf{x_i})$. Thus, with a Gaussian kernel, the mapped datapoints are distributed as points on a hypersphere of unit radius in feature space (since $\Phi(\mathbf{x}) \cdot \Phi(\mathbf{x}) = e^0 = 1$). If we set $\phi(\mathbf{z}) = \mathbf{w} \cdot \Phi(\mathbf{z}) + b = \sum_i \alpha_i K(\mathbf{z}, \mathbf{x}_i) + b = 0$, we define a hyperplane in feature space. The idea is to pull this hyperplane onto the mapped datapoints so that the margin is always zero or positive where the data is located. We pull the hyperplane onto the datapoints by minimizing $\sum_i \phi(\mathbf{x}_i)$. This is achieved by:

$$\min_{\alpha, b} \left\{ \sum_{i=1}^{m} \left(\sum_{j=1}^{m} \alpha_j K(\mathbf{x}_i, \mathbf{x}_j) + b \right) \right\} \qquad (2.17)$$

subject to:

$$\sum_{j=1}^{m} \alpha_j K(\mathbf{x}_i, \mathbf{x}_j) + b \geq 0 \qquad \forall i \qquad (2.18)$$

$$\sum_{i=1}^{m} \alpha_i = 1, \quad \alpha_i \geq 0 \qquad (2.19)$$

The bias b is just treated as an additional parameter in the minimization process, though unrestricted in sign. Many real-life datasets contain outliers. To handle these, we can introduce a *soft margin* and perform the following minimization instead:

$$\min_{\alpha, b, \xi} \left\{ \sum_{i=1}^{m} \left(\sum_{j=1}^{m} \alpha_j K(\mathbf{x}_i, \mathbf{x}_j) + b \right) + \lambda \sum_{i=1}^{m} \xi_i \right\} \qquad (2.20)$$

subject to (2.19) and:

$$\sum_{j=1}^{m} \alpha_j K(\mathbf{x}_i, \mathbf{x}_j) + b \geq -\xi_i, \quad \xi_i \geq 0. \qquad (2.21)$$

The parameter λ controls the extent of margin errors (larger λ means fewer outliers are ignored: $\lambda \to \infty$ corresponds to the *hard margin* limit).

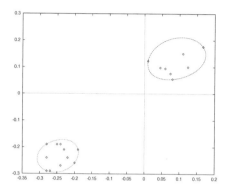
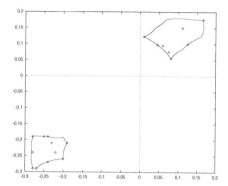

Figure 2.2: The boundary around two clusters of data in input space using the LP method in (2.17-2.19). *Left*: a hard margin solution using a Gaussian kernel. *Right*: a solution using the modified Gaussian kernel $K(\mathbf{x}_i, \mathbf{x}_j) = e^{-|\mathbf{x}_i - \mathbf{x}_j|/2\sigma^2}$.

In Figure 2.2, we see the boundary in input space placed around two clusters of data using a Gaussian and modified Gaussian kernel using this linear programming approach. The boundary encloses the data and passes through several datapoints, which we call *support objects*. However, this would mean a datapoint could be quite close to one of these support objects while still being classed as novel. The standard approach to this problem is to introduce *virtual datapoints*. Thus around each datapoint, we introduce a set virtual points through use of a Gaussian deviate (a random number generator, which creates points distributed as a Gaussian probability distribution). This will inflate the boundary while retaining its approximate shape.

Further reading: based on prior work (Burges [1998], Schölkopf et al. [1995]), the above QP approach to novelty detection was proposed by Tax and Duin [1999] and used by the latter authors on real life applications (Tax et al. [1999]). An alternative QP approach has been developed by Schölkopf et al. [1999] and Schölkopf and Smola [2002a]. Instead of *attracting* the hyperplane toward the datapoints, as in the above LP approach, this approach is based on *repelling* the hyperplane away from the origin while maintaining the requirement $\mathbf{w} \cdot \mathbf{x}_i + b \geq 0$. The LP approach in (2.17 - 2.19) was presented by Campbell and Bennett [2001].

2.5 REGRESSION: LEARNING WITH REAL-VALUED LABELS

So far, we have only considered learning with discrete labels. We now consider *regression* and the construction of models with *real-valued labels* y_i. We will introduce regression with one of the simplest

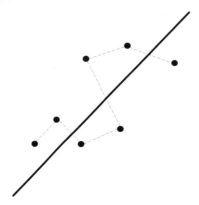

Figure 2.3: An illustration of *overfitting*: the seven points are assumed to be drawn from a distribution depicted by the (solid) line. The data is assumed corrupted by noise, so that these points lie on both sides of the line. An overfitting solution is depicted by the dashed line. This solution correctly passes through the seven datapoints. However, it is evidently a poor model of the underlying distribution (the straight line) and can be expected to result in inaccurate predictions.

methods which can be kernelized, namely *ridge regression*. To model the dependencies between the input vectors $\mathbf{x}_i$ and the y_i, we will use a linear function of the form:

$$g(\mathbf{x}_i) = \mathbf{w}^T \mathbf{x}_i \tag{2.22}$$

to approximate y_i. One method for finding appropriate weight values $\mathbf{w}$ is to minimize a *loss function* in $\mathbf{w}$. Here we use a quadratic error term:

$$L(\mathbf{w}) = \frac{1}{2} \sum_{i=1}^{m} (y_i - g(\mathbf{x}_i))^2 \tag{2.23}$$

If we did achieve a solution $y_i = g(\mathbf{x}_i)$, there is a risk that we may overfit the data leading to a solution exhibiting poor generalization (Figure 2.3). Mapping the data to feature space would potentially make overfitting worse since, in a higher dimensional space, there is greater flexibility and thus a much broader range of models capable of fitting the data.

The problem of overfitting is handled by introducing a *regularization term*, and one of the simplest ways to do this is to penalize $\mathbf{w}^T \mathbf{w}$, as for the primal formulation of an SVM classifier. Thus, in place of (2.23), we minimize the following loss function:

$$L(\mathbf{w}, \lambda) = \frac{1}{2} \sum_{i=1}^{m} \left(y_i - \mathbf{w}^T \mathbf{x}_i\right)^2 + \frac{1}{2}\lambda \mathbf{w}^T \mathbf{w} \tag{2.24}$$

If some components of $\mathbf{w}$ go to zero at the end of minimization then $g(\mathbf{x}_i)$ is independent of the corresponding features in $\mathbf{x}_i$, and model complexity has been reduced. At the minimum of (2.24) we get:

$$\frac{\partial L}{\partial \mathbf{w}} = -\sum_{i=1}^{m} \left(y_i - \mathbf{w}^T \mathbf{x}_i \right) \mathbf{x}_i + \lambda \mathbf{w} = 0 \qquad (2.25)$$

From which we get the solution:

$$\mathbf{w} = \left(\sum_{i=1}^{m} \mathbf{x}_i \mathbf{x}_i^T + \lambda \mathbf{I} \right)^{-1} \left(\sum_{j=1}^{m} \mathbf{x}_j y_j \right) \qquad (2.26)$$

We therefore see a second advantage from usage of the regularization term. With $\lambda = 0$ the matrix $\sum_{i=1}^{m} \mathbf{x}_i \mathbf{x}_i^T$ may not be invertable. However, with $\lambda \neq 0$, the inverse is always numerically stable. With a real-life application, λ is typically found by a cross-validation study.

We can now introduce kernel substitution through a mapping to feature space $\mathbf{x}_i \rightarrow \Phi(\mathbf{x}_i)$. Let $\mathbf{y}$ be a vector with m scalar components y_i and Φ a matrix *whose rows are the transpose of the mapping vectors* $\Phi(\mathbf{x}_i)$, then we can write (2.26) as:

$$\mathbf{w} = \left(\lambda \mathbf{I} + \Phi \Phi^T \right)^{-1} (\Phi \mathbf{y}) \qquad (2.27)$$

We can then use a result from matrix algebra, namely the Woodbury-Sherman-Morrison matrix inversion formula:

$$\left(\mathbf{A}^{-1} + \mathbf{B}^T \mathbf{C}^{-1} \mathbf{B} \right)^{-1} \mathbf{B}^T \mathbf{C}^{-1} = \mathbf{A} \mathbf{B}^T \left(\mathbf{B} \mathbf{A} \mathbf{B}^T + \mathbf{C} \right)^{-1} \qquad (2.28)$$

With $\mathbf{A}^{-1} = \lambda \mathbf{I}$, $\mathbf{B} = \Phi^T$ and $\mathbf{C} = \mathbf{I}$, this gives:

$$\mathbf{w} = \frac{1}{\lambda} \Phi \left(\frac{1}{\lambda} \Phi^T \Phi + \mathbf{I} \right)^{-1} \mathbf{y} = \Phi \left(\Phi^T \Phi + \lambda \mathbf{I} \right)^{-1} \mathbf{y} \qquad (2.29)$$

If we further introduce the notation:

$$\alpha = \left(\Phi^T \Phi + \lambda \mathbf{I} \right)^{-1} \mathbf{y} \qquad (2.30)$$

then we can write:

$$\mathbf{w} = \sum_{i=1}^{m} \alpha_i \Phi(\mathbf{x}_i) \qquad (2.31)$$

This is the analogue of the equation $\mathbf{w} = \sum_{i=1}^{m} \alpha_i y_i \Phi(\mathbf{x}_i)$ found for classification (see (1.6)), and it shows that, as for classification, the $\mathbf{w}$ can be written as a linear combination of the mapped datapoints $\Phi(\mathbf{x}_i)$. A further observation is that the $\Phi^T \Phi$ are the kernel matrices with components $K(\mathbf{x}_i, \mathbf{x}_j) = \Phi^T(\mathbf{x}_i)\Phi(\mathbf{x}_j)$. Let $\mathbf{K} = \Phi^T \Phi$ then:

$$\alpha = (\mathbf{K} + \lambda \mathbf{I})^{-1} \mathbf{y} \tag{2.32}$$

For a novel input $\mathbf{z}$, and writing $\overline{\mathbf{K}} = K(\mathbf{x}_i, \mathbf{z})$, we can make a prediction $\overline{y}$, based on use of the kernel matrix and the y_i only since:

$$\overline{y} = \mathbf{w}^T \Phi(\mathbf{z}) \doteq \mathbf{y} (\mathbf{K} + \lambda \mathbf{I})^{-1} \overline{\mathbf{K}} \tag{2.33}$$

The above formulation can also be derived from a primal formulation paralleling our derivation of the Support Vector Machine for classification. Thus we introduce a variable $\xi_i = y_i - \mathbf{w}^T \Phi(\mathbf{x}_i)$, which quantifies the error in (2.23). The primal formulation is a constrained QP problem:

$$\min_{\mathbf{w}, \xi} \left\{ L = \sum_{i=1}^m \xi_i^2 \right\} \tag{2.34}$$

subject to:

$$
\begin{aligned}
y_i - \mathbf{w}^T \Phi(\mathbf{x}_i) &= \xi_i & \forall i \\
\mathbf{w}^T \mathbf{w} &\leq B^2
\end{aligned}
$$

where the last constraint enforces regularization on $\mathbf{w}$. Introducing Lagrange multiplers β_i and λ for these two constraints, we get the Lagrange function:

$$L = \sum_{i=1}^m \xi_i^2 + \sum_{i=1}^m \beta_i \left[y_i - \mathbf{w}^T \Phi(\mathbf{x}_i) - \xi_i \right] + \lambda \left(\mathbf{w}^T \mathbf{w} - B^2 \right) \tag{2.35}$$

From the KKT conditions at the solution, we deduce that:

$$\xi_i = \frac{1}{2} \beta_i \qquad \mathbf{w} = \frac{1}{2\lambda} \sum_{i=1}^m \beta_i \Phi(\mathbf{x}_i) \tag{2.36}$$

Resubstituting these back into L, we obtain the dual:

$$W = \sum_{i=1}^m \left(-\frac{1}{4}\beta_i^2 + \beta_i y_i \right) - \frac{1}{4\lambda} \sum_{i,j=1}^m \left(\beta_i \beta_j K(\mathbf{x}_i, \mathbf{x}_j) \right) - \lambda B^2 \tag{2.37}$$

Let us introduce a new variable through a positive rescaling of β_i ($\lambda \geq 0$ since it is a Lagrange multiplier):

$$\alpha_i = \frac{1}{2\lambda} \beta_i \tag{2.38}$$

We then arrive at the following restatement of the dual:

$$\max_{\alpha_i, \lambda} \left\{ W = -\lambda^2 \sum_{i=1}^{m} \alpha_i^2 + 2\lambda \sum_{i=1}^{m} \alpha_i y_i - \lambda \sum_{i,j=1}^{m} \alpha_i \alpha_j K(\mathbf{x}_i, \mathbf{x}_j) - \lambda B^2 \right\} \tag{2.39}$$

From the derivative of W with respect to α_i, we find the same solution as in (2.29):

$$\alpha = (\mathbf{K} + \lambda \mathbf{I})^{-1} \mathbf{y} \tag{2.40}$$

Though ridge regression can be kernelized and offers a practical route to regression, it does not give a sparse representation in terms of the datapoints or samples. For prediction on a novel datapoint (2.33), we implicitly use *all* the datapoints via the kernel matrix $\mathbf{K}$. However, sample sparsity is desirable since it reduces the complexity of the model. The SVM classification solution is sparse since some datapoints are not support vectors and thus do not contribute to the decision function. This sample sparsity emerged from the constraints via the KKT condition $\alpha_i(y_i(\mathbf{w} \cdot \mathbf{x}_i + b) - 1) = 0$, which dictated $\alpha_i = 0$ for non-support vectors. The approach we now describe introduces this sample sparsity by a similar use of constraint conditions.

The following approach to regression is theoretically motivated by statistical learning theory. Instead of (1.3), we now use constraints $y_i - \mathbf{w}^T \mathbf{x}_i - b \leq \epsilon$ and $\mathbf{w}^T \mathbf{x}_i + b - y_i \leq \epsilon$ to allow for some deviation ϵ between the target for the outputs, the labels y_i, and the function $g(\mathbf{x}) = \mathbf{w}^T \mathbf{x} + b$. We can visualise this as a band or tube of size $\pm(\theta - \gamma)$ around the hypothesis function $g(\mathbf{x})$, and any points outside this tube can be viewed as training errors. The structure of the tube is defined by an $\epsilon-$*insensitive* loss function (Figure 2.4). As before, we minimize the sum of the squares of the weight-vectors $\|\mathbf{w}\|^2$ to penalise overcomplexity. To account for training errors, we also introduce slack variables $\xi_i, \widehat{\xi}_i$ for the two types of error. These slack variables are zero for points inside the tube and progressively increase for points outside the tube, according to the loss function used. This approach is called ϵ-*SV regression*. As for the soft margin parameters of Chapter 1, we can find the best value of ϵ as that value minimizing the quadratic loss function on validation data.

A linear $\epsilon-$insensitive loss function: For a *linear $\epsilon-$insensitive loss function* the task is to minimize:

$$\min_{\mathbf{w}, \xi_i, \widehat{\xi}_i} \left[\mathbf{w}^T \mathbf{w} + C \sum_{i=1}^{m} \left(\xi_i + \widehat{\xi}_i \right) \right] \tag{2.41}$$

subject to

$$\begin{aligned} y_i - \mathbf{w}^T \mathbf{x}_i - b &\leq \epsilon + \xi_i \\ (\mathbf{w}^T \mathbf{x}_i + b) - y_i &\leq \epsilon + \widehat{\xi}_i \end{aligned} \tag{2.42}$$

where the slack variables are both positive $\xi_i, \widehat{\xi}_i \geq 0$. After kernel substitution, we find the dual objective function:

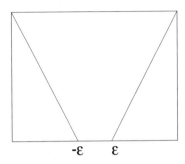 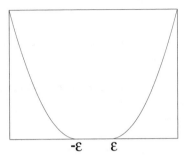

Figure 2.4: *Left*: a linear ϵ-insensitive loss function versus $y_i - \mathbf{w} \cdot \mathbf{x}_i - b$. *Right*: a quadratic ϵ-insensitive loss function.

$$
\begin{aligned}
W(\alpha, \widehat{\alpha}) = {} & \sum_{i=1}^{m} y_i(\alpha_i - \widehat{\alpha}_i) - \epsilon \sum_{i=1}^{m}(\alpha_i + \widehat{\alpha}_i) \\
& - \frac{1}{2} \sum_{i,j=1}^{m}(\alpha_i - \widehat{\alpha}_i)(\alpha_j - \widehat{\alpha}_j)K(x_i, x_j)
\end{aligned}
\tag{2.43}
$$

which is maximized subject to

$$
\sum_{i=1}^{m}\widehat{\alpha}_i = \sum_{i=1}^{m}\alpha_i
\tag{2.44}
$$

and:

$$
0 \leq \alpha_i \leq C \qquad 0 \leq \widehat{\alpha}_i \leq C
$$

A quadratic $\epsilon-$insensitive loss function: Similarly, a *quadratic $\epsilon-$insensitive loss function* gives rise to:

$$
\min_{\mathbf{w}, \xi_i, \widehat{\xi}_i}\left[\mathbf{w}^T\mathbf{w} + C\sum_{i=1}^{m}\left(\xi_i^2 + \widehat{\xi}_i^2\right)\right]
\tag{2.45}
$$

subject to (2.42), giving a dual objective function:

$$
\begin{aligned}
W(\alpha, \widehat{\alpha}) \;=\;& \sum_{i=1}^{m} y_i(\alpha_i - \widehat{\alpha}_i) - \epsilon \sum_{i=1}^{m} (\alpha_i + \widehat{\alpha}_i) \\
& - \frac{1}{2} \sum_{i,j=1}^{m} (\alpha_i - \widehat{\alpha}_i)(\alpha_j - \widehat{\alpha}_j)\left(K(\mathbf{x}_i, \mathbf{x}_j) + \delta_{ij}/C\right)
\end{aligned}
\tag{2.46}
$$

which is maximized subject to (2.44). In both cases the function modelling the data is then:

$$
g(\mathbf{z}) = \sum_{i=1}^{m} (\alpha_i^\star - \widehat{\alpha}_i^\star) K(\mathbf{x}_i, \mathbf{z}) + b^\star
\tag{2.47}
$$

The solution satisfies the KKT condition (see Chapter A.3) which, for a linear loss function, include the conditions:

$$
\begin{aligned}
\alpha_i \left(\epsilon + \xi_i - y_i + \mathbf{w}^T \mathbf{x}_i + b\right) &= 0 \\
\widehat{\alpha}_i \left(\epsilon + \widehat{\xi}_i + y_i - \mathbf{w}^T \mathbf{x}_i - b\right) &= 0
\end{aligned}
\tag{2.48}
$$

where $\mathbf{w} = \sum_{j=1}^{m} y_j(\alpha_j - \widehat{\alpha}_j)\mathbf{x}_j$, and:

$$
\begin{aligned}
(C - \alpha_i)\, \xi_i &= 0 \\
(C - \widehat{\alpha}_i)\, \widehat{\xi}_i &= 0
\end{aligned}
\tag{2.49}
$$

From the latter conditions, we see that only when $\alpha_i = C$ or $\widehat{\alpha}_i = C$ are the slack variables $(\xi_i, \widehat{\xi}_i)$ non-zero: these examples correspond to points outside the ϵ-insensitive tube.

We compute the bias, $b^\star$ from the KKT conditions. Thus from (2.48), we can find the bias from a *non-bound* example with $0 < \alpha_i < C$ using $b = y_i - \mathbf{w}^T \mathbf{x}_i - \epsilon$, and similarly for $0 < \widehat{\alpha}_i < C$, we can obtain it from $b = y_i - \mathbf{w}^T \mathbf{x}_i + \epsilon$. Though the bias can be obtained from one such example, it is best to compute it using an average over a set of non-bound examples.

ν-**SV Regression**: As for classification, where we introduced νSVM, we can also formulate regression to improve interpretability of the model parameters. Specifically, we use a parameter ν, which bounds the fraction of datapoints lying outside the tube. Using the same regression model as in (2.47), the dual formulation of ν-SV Regression is:

$$
\max_{\{\alpha, \widehat{\alpha}_i\}} \left\{ W(\alpha, \widehat{\alpha}) = \sum_{i=1}^{m} y_i\, (\alpha_i - \widehat{\alpha}_i) - \frac{1}{2} \sum_{i,j=1}^{m} (\alpha_i - \widehat{\alpha}_i)\left(\alpha_j - \widehat{\alpha}_j\right) K(\mathbf{x}_i, \mathbf{x}_j) \right\}
\tag{2.50}
$$

subject to:

$$\sum_{i=1}^{m} (\alpha_i - \widehat{\alpha}_i) \;=\; 0 \tag{2.51}$$

$$\sum_{i=1}^{m} (\alpha_i + \widehat{\alpha}_i) \;\leq\; \nu C \tag{2.52}$$

$$0 \;\leq\; \alpha_i \leq C/m \qquad 0 \leq \widehat{\alpha}_i \leq C/m \tag{2.53}$$

Then we can establish that νm is an upper bound on the number of datapoints falling outside the tube and νm is a lower bound on the number of support vectors, in this case, datapoints with non-zero α_i or $\widehat{\alpha}_i$ which are on or outside the tube. As in previous models, we find the bias b from the solution by selecting a non-bound example i with $0 < \alpha_i < C/m$: the KKT conditions then imply $(\xi_i, \widehat{\xi}_i)$ are zero and hence b is implicitly found from $(\mathbf{w} \cdot \mathbf{x}_i + b) - y_i = \epsilon$ and $y_i - (\mathbf{w} \cdot \mathbf{x}_i + b) = \epsilon$ where $\mathbf{w} = \sum_{j=1}^{m} y_j(\alpha_j - \widehat{\alpha}_j)\mathbf{x}_j$).

Further reading: Ridge regression was initially discussed by Hoerl and Kennard [1970] with the dual form later introduced by Saunders et al. [1998]. ϵ-SV regression was originally proposed by (Vapnik [1995], Vapnik [1998]). Apart from the formulations given here, it is possible to define other loss functions giving rise to different dual objective functions. ν-SV Regression was proposed by Schölkopf et al. [1998] with a further discussion in Schölkopf and Smola [2002b]. As for classification and novelty detection, it is also possible to formulate a *linear programming* approach to regression (Weston et al. [1998]). This latter approach minimizes the number of support vectors favouring sparse hypotheses with smooth functional approximations of the data. A tutorial on SV regression is given in Smola and Schölkopf [2004].

2.6   STRUCTURED OUTPUT LEARNING

So far, we have considered SVM learning with simple outputs: either a discrete output label for classification or a continuously-valued output for regression. However, for many real-world applications, we would like to consider more complex output structures. Thus, for a given multi-class classification task, there may exist dependencies between the class labels. Previous models, which assumed independent labels, are therefore inappropriate. An obvious context would be data which is organised into a taxonomy: some labelled data objects could be highly dependent if they are close together in the same sub-tree. These tasks are commonly referred to as *structured output prediction problems*. With structured output prediction, we wish to capture the dependency structure across the output class labels so as to generalize across classes. For typical applications, the size of the output space is very large, thus we need to devise specific methodologies to handle these types of problem.

As an explicit example, given an input sentence, we may want to output a syntax or parse tree which portrays the syntactic structure of the sentence. In Figure 2.5 we give an example in which a parse tree has been generated from the sentence 'Paul ate the pie'. The sentence (S) has been parsed into a noun phrase (NP) and a verb phrase (VB). 'Paul' is simply a noun (N) which is the subject of this sentence. The verb phrase consists of a verb ('ate') and a noun phrase which has a determiner (Det, in this case the definite article 'the') and a noun ('pie'). Each node in the parse tree corresponds to usage of a particular grammar rule, for example, S→ NP, VB. We may wish to generate output tree structures in many contexts, for example, if we are parsing a computer program into its component instructions or categorising the functional components, the genes, embedded in a genetic sequence.

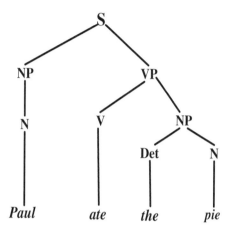

Figure 2.5:   A simple parse tree for the sentence 'Paul ate the pie'.

To proceed further, we need a function to encode the mapping from the input $\mathbf{x}$ to the given structured output $\mathbf{y}$. In the example given, $\mathbf{x}$ is the input sentence 'Paul ate the pie', and $\mathbf{y}$ is the resulting parse tree given in Figure 2.5. The function $\Psi(\mathbf{x}, \mathbf{y})$ encodes this mapping for the given vocabulary space and formal grammar used:

$$
\Psi(\mathbf{x}, \mathbf{y}) \;=\;
\begin{pmatrix}
1 \\ 0 \\ 1 \\ 1 \\ \vdots \\ 1 \\ 1 \\ 1 \\ 1
\end{pmatrix}
\begin{array}{l}
\text{S} \to \text{NP, VB} \\
\text{S} \to \text{NP} \\
\text{NP} \to \text{Det, N} \\
\text{VP} \to \text{V, NP} \\
\cdots \\
\text{N} \to \text{Paul} \\
\text{V} \to \text{ate} \\
\text{Det} \to \text{the} \\
\text{N} \to \text{pie}
\end{array}
$$

thus $\Psi(\mathbf{x}, \mathbf{y})$ is a vector whose components are the counts of how often a grammar rule occurs in the parse tree $\mathbf{y}$. We will associate a weight w_l to each node in the tree. This, in turn, means we can derive a predictive function given some novel input $\mathbf{z}$. Specifically, we can then state a function $F(\mathbf{z}, \mathbf{y}; \mathbf{w})$ as a weighted linear combination of the $\Psi(\mathbf{z}, \mathbf{y})$:

$$F(\mathbf{z}, \mathbf{y}; \mathbf{w}) = \mathbf{w}^T \Psi(\mathbf{z}, \mathbf{y}) \tag{2.54}$$

$F(\mathbf{z}, \mathbf{y}; \mathbf{w})$ can be regarded as a quantifying the compatibility of the pair $\mathbf{z}, \mathbf{y}$. The predicted structured output therefore amounts to maximizing this function over the space of parse trees $\mathcal{Y}$, thus:

$$f_w(\mathbf{z}) = \arg\max_{\mathbf{y} \in \mathcal{Y}} F(\mathbf{z}, \mathbf{y}; \mathbf{w}) \tag{2.55}$$

We still need to derive a scheme to train such a *structural* SVM, i.e., we need to find appropriate values for the weights $\mathbf{w}$ based on the use of training data consisting of paired data objects $(\mathbf{x}, \mathbf{y})$. To do so we further introduce a function $\triangle(\mathbf{y}, \widehat{\mathbf{y}})$ which quantifies the difference between the predicted output $\widehat{\mathbf{y}}$ and the correct output $\mathbf{y}$, given input $\mathbf{x}$. It is usual to consider functions $\triangle(\mathbf{y}, \mathbf{y}')$, for arbitrary output labels $\mathbf{y}$ and $\mathbf{y}'$ such that $\triangle(\mathbf{y}, \mathbf{y}) = 0$ and $\triangle(\mathbf{y}, \mathbf{y}') > 0$ if $\mathbf{y} \neq \mathbf{y}'$. Intuitively, the training process should involve the minimization of $\triangle(\mathbf{y}, \widehat{\mathbf{y}})$, so that $\widehat{\mathbf{y}}$ closely agrees with $\mathbf{y}$. However, in general, $\triangle$ is not a convex function, and it will have discontinuities. An example would be the choice $\triangle(\mathbf{y}, \mathbf{y}) = 0$, $\triangle(\mathbf{y}, \mathbf{y}') = 1$ if $\mathbf{y} \neq \mathbf{y}'$, which is discontinuous and does not give a unique solution. To circumvent this problem, we avoid minimizing $\triangle(\mathbf{y}, \widehat{\mathbf{y}})$ directly but instead minimize a more tractable upper bound on $\triangle(\mathbf{y}, \widehat{\mathbf{y}})$. Thus for a given input $\mathbf{x}_i$ with matching output $\mathbf{y}_i$ we use:

$$\triangle(\mathbf{y}_i, \widehat{\mathbf{y}}_i) \leq \max_{\mathbf{y} \in \mathcal{Y}} \left[\triangle(\mathbf{y}_i, \mathbf{y}) + \mathbf{w}^T \Psi(\mathbf{x}_i, \mathbf{y}) \right] - \mathbf{w}^T \Psi(\mathbf{x}_i, \mathbf{y}_i) \tag{2.56}$$

This bound follows by noting that:

$$
\begin{aligned}
\triangle(\mathbf{y}_i, \widehat{\mathbf{y}}_i) \; &\leq \; \triangle(\mathbf{y}_i, \widehat{\mathbf{y}}_i) - \left[\mathbf{w}^T \Psi(\mathbf{x}_i, \mathbf{y}_i) - \mathbf{w}^T \Psi(\mathbf{x}_i, \widehat{\mathbf{y}}) \right] \\
&= \; \max_{\mathbf{y} \in \mathcal{Y}} \left[\triangle(\mathbf{y}_i, \mathbf{y}) \right] - \mathbf{w}^T \Psi(\mathbf{x}_i, \mathbf{y}_i) + \max_{\mathbf{y} \in \mathcal{Y}} \left[\mathbf{w}^T \Psi(\mathbf{x}_i, \mathbf{y}) \right]
\end{aligned}
$$

since $\widehat{\mathbf{y}}$ maximizes $\mathbf{w}^T \Psi(\mathbf{x}_i, \mathbf{y})$. We further introduce a $||\mathbf{w}||^2$ regularizer to finally give the following optimization problem for a structural SVM:

$$W = \min_{\mathbf{w}} \left\{ \frac{1}{2} \mathbf{w}^T \mathbf{w} + C \sum_{i=1}^{m} \left[\max_{\mathbf{y} \in \mathcal{Y}} \left(\triangle(\mathbf{y}_i, \mathbf{y}) + \mathbf{w}^T \Psi(\mathbf{x}_i, \mathbf{y}) \right) - \mathbf{w}^T \Psi(\mathbf{x}_i, \mathbf{y}_i) \right] \right\} \tag{2.57}$$

We could rewrite this problem as:

$$
\begin{aligned}
W \; &= \; \min_{\mathbf{w}} \left\{ \left[\frac{1}{2} \mathbf{w}^T \mathbf{w} + C \sum_{i=1}^{m} \max_{\mathbf{y} \in \mathcal{Y}} \left(\triangle(\mathbf{y}_i, \mathbf{y}) + \mathbf{w}^T \Psi(\mathbf{x}_i, \mathbf{y}) \right) \right] - \left[C \sum_{i=1}^{m} \mathbf{w}^T \Psi(\mathbf{x}_i, \mathbf{y}_i) \right] \right\} \\
&= \; \min_{\mathbf{w}} \left\{ [f(\mathbf{w})] - [g(\mathbf{w})] \right\}
\end{aligned}
$$

which amounts to the difference of two convex functions. As such, it can be solved by a *difference of convex* (DC) optimization problem, which is guaranteed to minimize W at each iteration and which will converge to a local minimum (see Further Reading). An alternative approach is to introduce a slack variable for each sample ξ_i and reformulate it as a familiar constrained quadratic programming problem:

$$\min_{\mathbf{w}, \xi} \left[\frac{1}{2} \mathbf{w}^T \mathbf{w} + C \sum_{i=1}^{m} \xi_i \right] \tag{2.58}$$

subject to the set of constraints:

$$\mathbf{w}^T \Psi(\mathbf{x}_i, \mathbf{y}_i) - \mathbf{w}^T \Psi(\mathbf{x}_i, \widehat{\mathbf{y}}) + \xi_i \geq \triangle(\mathbf{y}_i, \widehat{\mathbf{y}}) \tag{2.59}$$

for all $\widehat{\mathbf{y}} \in \mathcal{Y}$, with $\xi_i \geq 0 \; \forall i$ and where $i = 1, \ldots, m$.

Further reading: There are many proposed approaches to structured output prediction. The above description follows the maximal margin approaches described in (Taskar et al. [2005], Tsochantaridis et al. [2005]). SVM structural learning is reviewed in Bakir et al. [2007]. Concave-convex procedures, which can handle difference of convex problems, are considered by Yuille and Rangarajan [2003].

CHAPTER 3

Learning with Kernels

3.1 INTRODUCTION

So far, we have used kernels which are fairly simple functions of the input data. However, real-life data can come in many different forms. Genetic sequences are encoded using nucleotides, protein sequences using amino acids. Graphs are also a useful means of portraying the relationship between data objects. In this chapter we will start with simple functional kernels. Thereafter, we consider kernels on strings and sequences and kernels *within* graphs and *between* graphs. The fact that we define kernels for many different types of input data illustrates the versatility of kernel-based methods but also points to a further use of kernel-based learning, namely, the ability to perform classification or regression based on multiple types of very distinct input data. This is called *multi-kernel learning* (MKL), and we end with an illustrative example.

3.2 PROPERTIES OF KERNELS

For any set of real-valued variables, $a_1, \ldots, a_m$, a *positive semi-definite kernel* (PSD kernel) satisfies:

$$\sum_{i=1}^{m} \sum_{j=1}^{m} a_i a_j K(\mathbf{x}_i, \mathbf{x}_j) \geq 0 \tag{3.1}$$

This type of kernel is symmetric, $K(\mathbf{x}_i, \mathbf{x}_j) = K(\mathbf{x}_j, \mathbf{x}_i)$, with positive components on the diagonal, $K(\mathbf{x}, \mathbf{x}) \geq 0$. An obvious example is the *linear kernel*:

$$K(\mathbf{x}_i, \mathbf{x}_j) = \mathbf{x}_i \cdot \mathbf{x}_j \tag{3.2}$$

This kernel is plainly symmetric and it satisfies (3.1) since:

$$\sum_{i=1}^{m} \sum_{j=1}^{m} a_i a_j (\mathbf{x}_i \cdot \mathbf{x}_j) = \left\| \sum_{i=1}^{m} a_i \mathbf{x}_i \right\|^2 \geq 0 \tag{3.3}$$

With a mapping into feature space we can also satisfy this positive semi-definite requirement. Thus suppose we use $\Phi(\mathbf{x})$ to perform a mapping into a d-dimensional space, which we denote $\mathbb{N}^d$, and let $(\mathbf{a} \cdot \mathbf{b})_{\mathbb{N}^d}$ denote an inner product between $\mathbf{a}$ and $\mathbf{b}$ in this space, then:

$$\sum_{i=1}^{m} \sum_{j=1}^{m} a_i a_j \left(\Phi(\mathbf{x}_i) \cdot \Phi(\mathbf{x}_j) \right)_{\mathbb{N}^d} = \left\| \sum_{i=1}^{m} a_i \Phi(\mathbf{x}_i) \right\|_{\mathbb{N}^d}^2 \geq 0 \tag{3.4}$$

Of course, this statement is only correct if an inner product can be defined in feature space. A *Hilbert space* is an inner product space, which is also complete with respect to the norm defined by the inner product. A PSD kernel K would therefore be permissible if and only if there exists a Hilbert space $\mathcal{H}$ and a mapping $\Phi : \mathcal{X} \to \mathcal{H}$ such that for any $\mathbf{x}_i$ and $\mathbf{x}_j$ in $\mathcal{X}$:

$$K(\mathbf{x}_i, \mathbf{x}_j) = \big(\Phi(\mathbf{x}_i) \cdot \Phi(\mathbf{x}_j)\big)_{\mathcal{H}} \tag{3.5}$$

In fact, this is always true for a symmetric positive semi-definite matrix since it can be diagonalised using an orthonormal basis of eigenvectors $(u_1, \ldots, u_m)$ with non-negative eigenvalues $\lambda_m \geq \ldots \geq \lambda_1 \geq 0$ so:

$$K(\mathbf{x}_i, \mathbf{x}_j) = \sum_{k=1}^{m} \lambda_k u_k(i) u_k(j) = \big(\Phi(\mathbf{x}_i) \cdot \Phi(\mathbf{x}_j)\big)_{\mathbb{N}^m} \tag{3.6}$$

The mapping function is then $\Phi(\mathbf{x}_i) = (\sqrt{\lambda_1} u_1(i), \ldots, \sqrt{\lambda_m} u_m(i))$. *Thus, if a proposed kernel matrix is positive semi-definite, then it is a permissible kernel.* We can determine if a kernel matrix is PSD by determining its spectrum of eigenvalues. Thus if a proposed kernel matrix has at least one negative eigenvalue λ with corresponding eigenvector $\mathbf{v}$, say, then $\mathbf{v}^T K \mathbf{v} = \lambda \mathbf{v}^T \mathbf{v} < 0$, so it is not PSD.

From the class of permissible kernels, we can construct other permissible kernels. For example:

[1] If $K_1(\mathbf{x}_i, \mathbf{x}_j)$ is a kernel, then so is:

$$K(\mathbf{x}_i, \mathbf{x}_j) = c K_1(\mathbf{x}_i, \mathbf{x}_j) \tag{3.7}$$

where c is a constant such that $c > 0$.

[2] If $K_1(\mathbf{x}_i, \mathbf{x}_j)$ and $K_2(\mathbf{x}_i, \mathbf{x}_j)$ are two kernels, then the sum $K(\mathbf{x}_i, \mathbf{x}_j) = K_1(\mathbf{x}_i, \mathbf{x}_j) + K_2(\mathbf{x}_i, \mathbf{x}_j)$ and the product $K(\mathbf{x}_i, \mathbf{x}_j) = K_1(\mathbf{x}_i, \mathbf{x}_j) K_2(\mathbf{x}_i, \mathbf{x}_j)$ are both permissible kernels: this will justify the multiple kernel learning approaches we introduce in Section 3.6.

[3] If $K_1(\mathbf{x}_i, \mathbf{x}_j)$ is a kernel and $f(\mathbf{x})$ is any function of $\mathbf{x}$, then the following is a permissible kernel:

$$K(\mathbf{x}_i, \mathbf{x}_j) = f(\mathbf{x}_i) K_1(\mathbf{x}_i, \mathbf{x}_j) f(\mathbf{x}_j) \tag{3.8}$$

[4] If $K_1(\mathbf{x}_i, \mathbf{x}_j)$ is a kernel, then so is:

$$K(\mathbf{x}_i, \mathbf{x}_j) = \exp(K_1(\mathbf{x}_i, \mathbf{x}_j)) \tag{3.9}$$

Since $\exp(\cdot)$ can be expanded in a series representation, this is a sub-instance of:

$$K(\mathbf{x}_i, \mathbf{x}_j) = p\left[K_1(\mathbf{x}_i, \mathbf{x}_j)\right] \tag{3.10}$$

where $p(\cdot)$ is a polynomial with non-negative coefficients.

A further manipulation we can apply to a kernel matrix is *normalization*. This is achieved using a mapping $\mathbf{x} \to \Phi(\mathbf{x}) / \|\Phi(\mathbf{x})\|_2$. The normalized kernel is then:

$$\widehat{K}(\mathbf{x}_i, \mathbf{x}_j) = \frac{\Phi(\mathbf{x}_i) \cdot \Phi(\mathbf{x}_j)}{\|\Phi(\mathbf{x}_i)\|_2 \|\Phi(\mathbf{x}_j)\|_2} = \frac{\Phi(\mathbf{x}_i) \cdot \Phi(\mathbf{x}_j)}{\sqrt{\Phi(\mathbf{x}_i) \cdot \Phi(\mathbf{x}_i)} \sqrt{\Phi(\mathbf{x}_j) \cdot \Phi(\mathbf{x}_j)}} \tag{3.11}$$

$$= \frac{K(\mathbf{x}_i, \mathbf{x}_j)}{\sqrt{K(\mathbf{x}_i, \mathbf{x}_i) K(\mathbf{x}_j, \mathbf{x}_j)}} \tag{3.12}$$

For the string and sequence kernels considered later, normalization of a kernel is useful for gaining document length independence. The normalized kernel also gives an insight into the Gaussian kernel, introduced earlier (1.10) since:

$$K(\mathbf{x}_i, \mathbf{x}_j) = \exp\left(-\frac{(\mathbf{x}_i - \mathbf{x}_j)^2}{2\sigma^2}\right) = \exp\left(\frac{\mathbf{x}_i \cdot \mathbf{x}_j}{\sigma^2} - \frac{\mathbf{x}_i \cdot \mathbf{x}_i}{2\sigma^2} - \frac{\mathbf{x}_j \cdot \mathbf{x}_j}{2\sigma^2}\right) \tag{3.13}$$

$$= \frac{\exp\left(\frac{\mathbf{x}_i \cdot \mathbf{x}_j}{\sigma^2}\right)}{\sqrt{\exp\left(\frac{\mathbf{x}_i \cdot \mathbf{x}_i}{\sigma^2}\right) \exp\left(\frac{\mathbf{x}_j \cdot \mathbf{x}_j}{\sigma^2}\right)}} \tag{3.14}$$

So the Gaussian kernel is a normalised kernel. Since we know that the linear kernel $K(\mathbf{x}_i, \mathbf{x}_j) = \mathbf{x}_i \cdot \mathbf{x}_j$ is valid, validity of a Gaussian kernel follows from properties [1], [3] and [4].

The class of useable kernels can be extended beyond those which are positive semi-definite. Thus a possible mapping function may not be positive semi-definite in general but could be permissible if it is positive definite with respect to the training data. If the kernel-based method is formulated directly in feature space (e.g., the LP method in Section 2.2), then the kernel matrix need not be PSD. Then again, we can handle non-PSD kernels (*indefinite kernels*) by the use of PSD *proxy kernels*.

Further Reading: Various approaches have been proposed for handling indefinite kernels. One method is to convert the indefinite kernel matrix into a PSD matrix by neglecting negative eigenvalues, using the absolute value of all eigenvalues or shifting eigenvalues to positivity by addition of a positive constant (cf. Graepel et al. [1998], Pekalska et al. [2002], Roth et al. [2003]). A further strategy is to derive a PSD proxy kernel (Luss and d'Aspremont [2008], Ying et al. [2009a]).

3.3   SIMPLE KERNELS

With many applications, simply defined kernels are sufficient. Generalizing the linear kernel, we have the *homogeneous polynomial kernel*:

$$K(\mathbf{x}_i, \mathbf{x}_j) = \left(\mathbf{x}_i \cdot \mathbf{x}_j\right)^d \tag{3.15}$$

and the *inhomogeneous polynomial*:

$$K(\mathbf{x}_i, \mathbf{x}_j) = \left(\mathbf{x}_i \cdot \mathbf{x}_j + c\right)^d \tag{3.16}$$

A common choice is the *Gaussian kernel*:

$$K(\mathbf{x}_i, \mathbf{x}_j) = \exp\left(-\frac{(\mathbf{x}_i - \mathbf{x}_j)^2}{2\sigma^2}\right) \tag{3.17}$$

We observe that $K(\mathbf{x}, \mathbf{x}) = 1$ for Gaussian kernels, so the mapped datapoints are located on a hypersphere of unit radius. Another choice is the *hyperbolic tangent kernel*:

$$K(\mathbf{x}_i, \mathbf{x}_j) = \tanh\left(\kappa(\mathbf{x}_i \cdot \mathbf{x}_j) + c\right) \tag{3.18}$$

where $\kappa > 0$ and $c < 0$. This kernel is not positive definite, in general, but has been used successfully in applications.

For each of these kernels, we notice that there is at least one *kernel parameter* (e.g., σ for the Gaussian kernel) whose value must be found. There are several ways to find the value of this parameter:

- If we have enough data, we can split it into a *training set*, a *validation set* and a *test set*. We then pursue a *validation* study in which we train the learning machine with different, regularly spaced, choices of the kernel parameter and use that value which gives the lowest error on the validation data (the validation data is therefore part of the training process). This is the preferred method if the data is available.

- If there is insufficient data for a validation dataset then we can use various criteria which estimate a test error from the training data: we describe some of these methods below for the case of binary classification.

- Other methods are possible. In Section 3.6, we discuss multiple kernel learning in which we learn the coefficients of a linear combination of kernels. Though frequently used with different types of kernels, we can also use multiple kernel learning with a linear combination of kernels, all of the same type (e.g., Gaussian kernels) but with each kernel having a different kernel parameter. Higher values for the kernel coefficients indicate more suitable values for the associated kernel parameters.

Finding the kernel parameter for classification: The test error can be estimated as a function of the kernel parameter, without recourse to validation data. Various criteria have been proposed, and here we briefly describe two fairly straightforward approaches for this purpose, for the case of binary classification.

From statistical learning theory, we can deduce an upper bound on the generalization error of an SVM, namely, $R^2/m\gamma^2$ where R is the radius of the smallest ball containing the training data and γ is the margin. At an optimum of the SVM dual objective function (1.12), we can deduce that $\gamma^2 = 1/\sum_{i=1}^m \alpha_i^\star$. This follows from the conditions $\sum_{i=1}^m \alpha_i^\star y_i = 0$, $\mathbf{w}^\star = \sum_{i=1}^m y_i \alpha_i^\star \mathbf{x}_i$ and $y_i(\mathbf{w}^\star \cdot \mathbf{x}_i + b^\star) = 1$, which apply at the optimum and so:

$$\frac{1}{\gamma^2} = \mathbf{w}^\star \cdot \mathbf{w}^\star = \sum_{i=1}^m y_i \alpha_i^\star \left(\mathbf{w}^\star \cdot \mathbf{x}_i\right) = \sum_{i=1}^m y_i \alpha_i^\star \left(\mathbf{w}^\star \cdot \mathbf{x}_i + b^\star\right) = \sum_{i=1}^m \alpha_i^\star \qquad (3.19)$$

In our discussion of one class classification (Section 2.4), we have already seen how we can calculate R, namely by solving the convex QP problem:

$$\max_\lambda \left\{ \sum_{i=1}^m \lambda_i K(\mathbf{x}_i, \mathbf{x}_j) - \sum_{i,j=1}^m \lambda_i \lambda_j K(\mathbf{x}_i, \mathbf{x}_j) \right\} \qquad (3.20)$$

such that $\sum_i \lambda_i = 1$ and $\lambda_i \geq 0$. The radius is then given by:

$$R^2 = \max_i \left\{ \sum_{j,k=1}^m \lambda_j \lambda_k K(\mathbf{x}_j, \mathbf{x}_k) - 2\sum_{j=1}^m \lambda_j K(\mathbf{x}_i, \mathbf{x}_j) + K(\mathbf{x}_i, \mathbf{x}_i) \right\} \qquad (3.21)$$

For Gaussian kernels, the datapoints lie on the surface of a hypersphere of unit radius since $\Phi(\mathbf{x}) \cdot \Phi(\mathbf{x}) = K(\mathbf{x}, \mathbf{x}) = 1$. So we can use the approximation $R \simeq 1$, and the bound is just $\sum_{i=1}^m \alpha_i^\star / m$. Thus, based on using the training set only, we could estimate the best value for the kernel parameter by sequentially training SVMs at regularly spaced values of the kernel parameter, evaluating the bound from the solution $\alpha_i^\star$, and choosing that value of the kernel parameter for which the bound is minimized. This method can work well in certain cases, e.g., for a Gaussian kernel with the datapoints spread evenly over the surface of the hypersphere. However, more refined estimates take into account the distribution of the data.

As a better criterion, we next consider a *leave-one-out bound*. For an SVM, this generalization bound is calculated by extracting a single datapoint and estimating the prediction error based on the hypothesis derived from the remaining $(m-1)$ datapoints. This calculation includes the use of a soft margin such as the box constraint $0 \leq \alpha_i \leq C$. Specifically, after finding the solution $(\alpha^\star, b^\star)$, the number of leave-one-out errors of an L_1-norm soft margin SVM is upper bounded by

$$|\{i : (2\alpha_i^\star B^2 + \xi_i^\star) \geq 1\}|/m \qquad (3.22)$$

that is, the count of the number of samples satisfying $2\alpha_i^\star B^2 + \xi_i \geq 1$ divided by the total number of samples, m. B^2 is an upper bound on $K(\mathbf{x}_i, \mathbf{x}_i)$ with $K(\mathbf{x}_i, \mathbf{x}_j) \geq 0$. We can determine a margin error $\xi_i^\star$ from $y_i(\sum_j \alpha_j^\star K(\mathbf{x}_j, \mathbf{x}_i) + b^\star) \geq 1 - \xi_i^\star$. For a Gaussian kernel $B = 1$ and with a hard margin this bound simply becomes the count of the number of instances i for which $2\alpha_i^\star \geq 1$, divided by the number of samples m. As above, this bound is estimated from the solution values $\alpha^\star$ at regularly spaced values of the kernel parameter with the value minimizing the bound retained for subsequent use.

Further reading: The first of these bounds was presented in Cristianini et al. [1999] and the leave-one-out bound in Joachims [2000]. Other rules include the *span-rule* (Chapelle and Vapnik [2000], Vapnik and Chapelle [2000]) and Bayesian approaches (Sollich [2002]).

3.4   KERNELS FOR STRINGS AND SEQUENCES

Strings are ordered sets of symbols drawn from an alphabet. They appear in many contexts. For example, we could be considering words from English text or strings of the four DNA-bases A, C, G and T, which make up DNA genetic sequences. We can readily observe a degree of similarity between strings. Thus suppose we consider the set of words *STRAY*, *RAY* and *RAYS*. They have the string *RAY* in common and a matching algorithm should pick up this similarity irrespective of the differing prefixes and suffixes (we are ignoring any similarities in meaning here and in our subsequent discussion). Strings could differ through deletions or insertions, thus *STAY* differs from *STRAY* by a gap consisting of a single deletion.

We can consider two distinct categories when matching ordered sets of symbols. The first we will call *string matching*: in this case contiguity of the symbols is important. For the second category, *sequence matching*, only the order is important. Thus for our example with *STAY* and *STRAY*, there are only two short contiguous strings in common *ST* and *AY*. On the other hand, *S*,*T*,*A* and *Y* are all ordered the same way in both words. Sequence kernels are important in many contexts. For example, when matching genetic sequences, there can often be extensive commonality between sequences interrupted by occasional deletions and insertions. In this Section, we consider three types of kernels. For the first, the *p*-spectrum kernel, contiguity is necessary. For the second, the subsequence kernel, contiguity is not necessary, and the order of symbols is important. Finally, we consider the gap-weighted kernel in which any non-contiguity is penalized.

The *p*-spectrum kernel: One way of comparing two strings is to count the number of contiguous substrings of length p which are in common between them. A *p*-spectrum kernel is based on the spectrum of frequencies of all *contiguous substrings of length p*. For example, suppose we wish to compute the 2-spectrum of the string $s = SAY$. There are two contiguous substrings of length $p = 2$, namely $u_1 = SA$ and $u_2 = AY$ both with a frequency of 1. As a further example let us consider the set of strings $s_1 = SAY$, $s_2 = BAY$, $s_3 = SAD$ and $s_4 = BAD$. The 2-spectra are given in the table below:

Table 3.1: A mapping function Φ for the *p*-spectrum kernel.

Φ	SA	BA	AY	AD
SAY	1	0	1	0
BAY	0	1	1	0
SAD	1	0	0	1
BAD	0	1	0	1

where each entry is the number of occurrences of the substring u (say $u = SA$) in the given string (say $s = SAY$). The corresponding kernel is then:

Table 3.2: The p-spectrum kernel matrix from the mapping function in Table 3.1.

K	SAY	BAY	SAD	BAD
SAY	2	1	1	0
BAY	1	2	0	1
SAD	1	0	2	1
BAD	0	1	1	2

Thus to compute the (SAY, BAY) entry in the kernel matrix, we sum the products of the corresponding row entries under each column in the mapping function Table 3.1. Only the pair of entries in the AY substring column both have non-zero entries of 1 giving $K(SAY, BAY) = 1$. For an entry on the diagonal of the kernel matrix, we take the sum of the squares of the entries in the corresponding row in Table 3.1. Thus, for $K(SAY, SAY)$, there are non-zero entries of 1 under SA and AY and so $K(SAY, SAY) = 2$.

Formally, we can define the mapping function and kernel as follows. Let $u \in \Sigma^p$ denote a arbitrary substring u of length p with symbols drawn from the alphabet Σ^p, and let v_1 and v_2 be strings (which can be empty), then the mapping amounts to the frequency count of the occurrence of u in s which is written:

$$\Phi_u^p(s) = |\{(v_1, v_2) : s = v_1 u v_2\}|. \qquad u \in \Sigma^p \qquad (3.23)$$

In our $p = 2$ example, for $u = AY$ and $s = SAY$, the only non-zero contribution has $v_1 = S$, v_2 is the empty set and the frequency count is 1. The p-spectrum kernel between strings s and t is thus:

$$K_p(s, t) = \sum_{u \in \Sigma^p} \Phi_u^p(s) \Phi_u^p(t) \qquad (3.24)$$

The all-subsequence kernel: The all-subsequences kernel uses a mapping function *over all contiguous and non-contiguous ordered subsequences of a string*, inclusive of the empty set. As an explicit example let us consider a two word example, consisting of $s_1 = SAY$ and $s_2 = BAY$. Let Ω represent the empty set, then the mapping is given in the table below:

Table 3.3: A mapping function Φ for the all-subsequences kernel.

Φ	Ω	A	B	S	Y	SA	BA	AY	SY	BY	SAY	BAY
SAY	1	1	0	1	1	1	0	1	1	0	1	0
BAY	1	1	1	0	1	0	1	1	0	1	0	1

The off-diagonal terms in the kernel matrix are then evaluated as the sum across all columns of the products of the two entries in each column. The diagonal terms are the sum of squares of all entries in a row:

Table 3.4: The all-subsequences kernel matrix for the mapping function in Table 3.3.

K	SAY	BAY
SAY	8	4
BAY	4	8

With *fixed length subsequence kernels,* we only consider non-contiguous subsequences with a fixed length p.

The gap-weighted subsequence kernel: with this kernel, positional information is included. A penalization is used so that intervening gaps or insertions decrease the score for the match. As an example, *SAY* is a subsequence of *STAY* and *SATURDAY*. However, *SAY* differs from *STAY* by one deletion, but *SAY* differs from *SATURDAY* by a gap of five 5 symbol deletions. Intuitively, the latter is more dissimilar from *SAY*. For a gap-weighted kernel of order p, the mapping function $\Phi_u^p(s)$ is the sum of the *weights* of the occurrences of a subsequence u of length p as a non-contiguous subsequence of s. The weight is of the form $\lambda^{length(u)}$. It is important to note that the subsequence can include other matching symbols.

As an example, let us evaluate $\Phi_u^p(s)$ for the subsequence $u = BAD$ in $s = CCBCADCB$. In this case $p = 3$ and $\Phi_u^p(s) = \lambda^4$. This is because $BCAD$ is the non-contiguous subsequence of s which contains the subsequence in u. Its length $length(u)$ is therefore 4. As a second example, let us find $\Phi_u^p(s)$ for the subsequence $u = BAD$ in $s = DAABCCACABDCCDB$. Again, $p = 3$ and, in this case, $\Phi_u^p(s) = 2\lambda^8 + 2\lambda^{11}$. The two subsequences of length 8 are both contained in the subsequence $BCCACABD$ (there are two matching symbols A), and the other two subsequences of length 11 are contained in the subsequence $BCCACABDCCD$. Having computed the mapping function, we then compute the kernel matrix in the usual way:

$$K_p(s, t) = \sum_{u \in \Sigma^p} \Phi_u^p(s)\Phi_u^p(t)$$

Returning to our example with the words *SAY*, *BAY*, *SAD* and *BAD*, we can compute the mapping function for $p = 2$ as:

Table 3.5: A mapping function for a $p = 2$ gap-weighted kernel.

Φ	SA	SY	AY	BA	BY	SD	AD	BD
SAY	λ^2	λ^3	λ^2	0	0	0	0	0
BAY	0	0	λ^2	λ^2	λ^3	0	0	0
SAD	λ^2	0	0	0	0	λ^3	λ^2	0
BAD	0	0	0	λ^2	0	0	λ^2	λ^3

Thus for the kernel component for $s = SAY$ and $t = BAY$, we add the products of the terms in the respective rows for SAY and BAY, giving:

$$K(SAY, BAY) = \lambda^4 \tag{3.25}$$

The diagonal terms in the kernel matrix are the sums of the squares of all entries in the corresponding row. Thus, for example:

$$K(SAY, SAY) = 2\lambda^4 + \lambda^6 \tag{3.26}$$

The normalized kernel $\widehat{K}(SAY, BAY)$ is then given by:

$$\widehat{K}(SAY, BAY) = \frac{K(SAY, BAY)}{\sqrt{K(SAY, SAY)K(BAY, BAY)}} = \frac{1}{(2 + \lambda^2)} \tag{3.27}$$

λ acts as a decay parameter which down-weights the matching score as sequences diverge due to insertions and deletions. As such the gap-weighted kernel interpolates between the above spectrum and fixed length subsequence kernels with the limit $\lambda \to 0$ corresponding to a spectrum kernel and $\lambda = 1$ corresponding to the latter. Thus, for example, if $\lambda \to 0$, then $\widehat{K}(SAY, BAY) = 1/2$ in (3.27). This is equal to the corresponding *normalized* component of the spectrum kernel matrix in Table 3.2.

Further Reading: for the computation of the p-spectrum kernel, for example, the number for substrings scales proportionally to the number of symbols in a string and comparison of a p-length substring involves p operations. If $|s|$ and $|t|$ are the lengths of strings s and t, we infer that a naive computation of the spectrum kernel has a computational cost $O(p\,|s|\,|t|)$. Consequently, a number of efficient routines have been proposed for computing string and sequence kernels. The first of these involves a *dynamic programming* approach (Lodhi et al. [2002]). In this case, we use a recursive approach with a base kernel (e.g., $K(s, \Omega) = 1$: the kernel for a string with the empty set is 1, see Table 3.3 and a recursive formula, which relates an unknown component of the kernel matrix to previously computed kernel matrix components. A second approach is based on *tries* or retrieval trees. This efficient construction reduces the cost of computing the p-spectrum kernel to $O(p(|s| + |t|))$ (Leslie and Kuang [2003]). A third approach is based on *suffix trees* (Vishwanathan and Smola [2003]). Shawe-Taylor and Cristianini [2004] give a comprehensive overview of a variety of kernels for handling strings, sequences, trees and text. Sonnenburg et al. [2007] discuss efficient methods for the fast learning of string kernels.

3.5 KERNELS FOR GRAPHS

Graphs are a natural way to represent many types of data. The vertices, or nodes, represent data objects and the edges would then represent the relationships between these data objects. With graphs, we can consider two types of similarity. For a given graph, we may be interested in the similarity of two vertices within the same graph. This graph could encode information about the similarities of datapoints represented as vertices. Alternatively, we may be interested in constructing a measure of similarity between two different graphs. We can construct kernels for both these similarity measures, and we refer to *kernels on graphs* when constructing a within-graph kernel between vertices and *kernels between graphs* for the latter comparison. For both types of comparison, the corresponding kernels must satisfy the requirements of being symmetric positive semi-definite.

Kernels on graphs: Let us first consider the construction of *kernels on graphs* for undirected graphs $G = (V, E)$ where V is the set of vertices and E are the edges. We will formulate the corresponding kernel as an *exponential kernel*. Specifically, the exponential of a $(n \times n)$ matrix βH can be written as a series expansion:

$$e^{\beta H} = I + \beta H + \frac{\beta^2}{2!}H^2 + \frac{\beta^3}{3!}H^3 + \dots \qquad (3.28)$$

where β is a real-valued scalar parameter (we comment on its interpretation later). Any even power of a symmetric matrix is positive semi-definite. Thus if H is symmetric and an even power, the exponential of this matrix is positive semi-definite and thus a suitable kernel. For our undirected graph with vertex set V and edge set E, we can use the notation $v_1 \sim v_2$ to denote a link or edge between vertices v_1 and v_2. Furthermore, let d_i denote the number of edges leading into vertex i. We can then construct a kernel using:

$$H_{ij} = \begin{cases} 1 & \text{for } i \sim j \\ -d_i & \text{for } i = j \\ 0 & \text{otherwise} \end{cases} \qquad (3.29)$$

To construct the exponential kernel representation, we then perform a *eigen-decomposition* of H. Any real symmetric matrix, such as H, can be diagonalised in the form $H = UDU^T$ where D is the diagonal matrix of ordered eigenvalues of H such that $\lambda_1 \geq \lambda_2 \geq \dots \geq \lambda_n$, and the columns of U are the corresponding orthonormal eigenvectors of H. Indeed, because $U^T U = I$, this result is generalizable to $H^k = UD^kU^T$, and through the above series, expansion representation to $e^{\beta H} =$

$U e^{\beta D} U^T$. Let us write the diagonal matrix as $D = diag(d_1, d_2, \ldots, d_n)$; the kernel is then given as:

$$
K = e^{\beta H} = U \begin{pmatrix} e^{\beta d_1} & 0 & \cdots & 0 \\ 0 & e^{\beta d_2} & \cdots & 0 \\ \vdots & \vdots & \ddots & \vdots \\ 0 & 0 & \cdots & e^{\beta d_n} \end{pmatrix} U^T \tag{3.30}
$$

This construction also applies to weighted symmetric graphs since for off-diagonal elements of H (i.e., $H_{ij}, i \neq j$), we would use the weight of the edge between i and j and similarly reweight the diagonal terms in H. In the above construction, the matrix H encodes *local* information about the direct linkages of vertices to each other whereas K encodes a *global* similarity measure over the entire graph G.

We additionally note that our kernel $K = e^{\beta H}$ satisfies the differential equation:

$$
\frac{d}{d\beta} K(\beta) = H K(\beta) \tag{3.31}
$$

This equates to a *diffusion equation* from physics, and for this reason, the above kernel is called a *diffusion kernel*. Diffusion equations describe the diffusion of heat or chemical concentrations through continuous media, and with this interpretation, we can view β as a parameter controlling extent of diffusion. Thus $\beta = 0$ would imply $K(\beta) = I$, and so no diffusion would occur. Diffusion is frequently modelled by random walks. If v_1 is a starting vertex, then a random walk is described by the path $(v_1, v_2, \ldots v_T)$ where v_{i+1} must be a neighbour of v_i $((v_i, v_{i+1}) \in E)$, and we pick v_{i+1} at random from the set of neighbours of v_i. It we let p_t be the probability of finding the random walker at a vertex at step t, then be can define a transition matrix Q such that $p_{t+1} = Q p_t$. Unfortunately, Q is not positive semi-definite and thus not suitable as a kernel. However, we can define a limiting case of an infinite number of random walk steps. With each iteration, there is an infinitesimally small probability of changing the current position: thus the transition matrix Q is very close to I, and we can state a kernel:

$$
K = e^{\beta H} = \lim_{n \to \infty} \left(I + \frac{\beta H}{n} \right)^n \tag{3.32}
$$

Expanding this expression into a power series gives the exponential kernel stated above.

Kernels between graphs: We may also want to derive a measure of similarity *between two graphs*, and the idea of walks on graphs will be useful here, too. Indeed, one approach to comparing two graphs is to compare walks on these two graphs. To simplify detail, we will consider *attributed* graphs whereby labels are attached to vertices and edges. We can define an adjacency matrix A on a graph G with components A_{ij} such that $A_{ij} = 1$ if $(v_i, v_j) \in E$ and 0 otherwise. Walks of length L can be found by talking the L'th power of this adjacency matrix. If we find $A^L(i, j) = k$, then k walks of length L exist between vertices i and j. To enumerate common walks on two different graphs, say $G = (V, E)$ and $\widetilde{G} = (\widetilde{V}, \widetilde{E})$, we can derive a *product graph*. Let the product graph be denoted $G_p = (V_p, E_p)$. Then G_p consists of pairs of identically labelled vertices and edges from the two graphs G and $\widetilde{G}$. Then if A_p is the adjacency matrix on G_p, common walks can be computed from A_p^L. We can then define a random walk kernel:

$$K_p(G, \widetilde{G}) = \sum_{i,j}^{|V_p|} \left[\sum_{n=0}^{\infty} \lambda^n A_p^n \right]_{ij} \tag{3.33}$$

where λ is a decay constant to ensure the summation converges. Conceptually, the random walk kernel compares two different kernels by counting the number of matching walks.

As for the computation of kernels on graphs, the evaluation of kernels between graphs is potentially computationally intensive. Thus if graphs G and $\widetilde{G}$ have the same number of labelled nodes n, computational comparison of all edges between G and $\widetilde{G}$ requires $O(n^4)$ steps. Evaluating powers of the adjacency matrix is even more compute intensive, requiring $O(n^6)$ steps. As the more dominant factor, the latter gives a runtime cost of $O(n^6)$. We can fortunately reduce the computational cost by a restatement involving Sylvester equations that is, if matrices A, B and C are known, finding X in the equation $X = AXB + C$. This reduces computation of the kernel to the more tractable runtime cost of $O(n^3)$.

Further reading: *Kernels on graphs* were initially proposed by Kondor and Lafferty [2002], who introduced the *diffusion kernel*, and later by Smola and Kondor [2003]. *Kernels between graphs* were initially proposed by Gartner et al. [2003] who considered random walk graph kernels. Applications to chemoinformatics (Kashima et al. [2004]) and protein function prediction (Borgwardt et al. [2005]) followed. The efficient computation of between-graph kernels using Sylvester's equation was proposed by Vishwanathan et al. [2008].

3.6   MULTIPLE KERNEL LEARNING

The performance of a SVM depends on the data representation via the choice of kernel function. In preceding chapters, we assumed the form of the kernel matrix was known *a priori* and left its choice to the practitioner. In this Section, we turn our attention to the problem of how to learn kernels automatically from the training data.

The *learning the kernel problem* ranges from finding the width parameter, σ, in a Gaussian kernel to obtaining an optimal linear combination of a finite set of candidate kernels, with these kernels representing different types of input data. The latter is referred to as *multiple kernel learning* (MKL). Specifically, let $\mathcal{K}$ be a set of candidate kernels, the objective of MKL is to design classification models which can simultaneously find both the optimal kernel and the best classifier. The typical choice for $\mathcal{K}$ is a linear combination of p prescribed kernels $\{K_\ell : \ell = 1, \ldots, p\}$ i.e.,

$$K = \sum_{\ell=1}^{p} \lambda_\ell K_\ell \tag{3.34}$$

where $\sum_{\ell=1}^{p} \lambda_\ell = 1$, $\lambda_\ell \geq 0$ and the λ_ℓ are called *kernel coefficients*. This framework is motivated by classification or regression problems in which we want to use all available input data, and this input data may derive from many disparate sources. These data objects could include network graphs and sequence strings in addition to numerical data: as we have seen that all these types of data can be encoded into kernels. The problem of data integration is therefore transformed into the problem of learning the most appropriate combination of candidate kernel matrices (typically, a linear combination is used though nonlinear combinations are feasible).

Further reading: Lanckriet et al. [2004a] pioneered work on MKL and proposed a semi-definite programming (SDP) approach to automatically learn a linear combination of candidate kernels for the case of SVMs. The further development of MKL included efficient optimization algorithms (Bach et al. [2004], Lanckriet et al. [2004a], Rakotomamonjy et al. [2008], Sonnenburg et al. [2006]) and the theoretical foundations of MKL (Bach [2008], Lanckriet et al. [2004a], Micchelli and Pontil [2005], Ying and Campbell [2009], Ying and Zhou [2007]).

3.7   LEARNING KERNEL COMBINATIONS VIA A MAXIMUM MARGIN APPROACH

There are a number of criteria for learning the kernel. We will consider the problem of how to find an optimal kernel for SVM binary classification. A good criterion would be to maximize the margin with respect to all possible kernel spaces capable of discriminating different classes of labeled data. In particular, for a given kernel $K \in \mathcal{K}$ with $K(\mathbf{x}_i, \mathbf{x}_j) = \Phi(\mathbf{x}_i) \cdot \Phi(\mathbf{x}_j)$, we have seen that maximizing the margin involves minimising $|\mathbf{w}^\star\|^2$ in feature space:

$$(\mathbf{w}^\star, b^\star) = \arg\min_{\mathbf{w}, b} \|\mathbf{w}\|^2 \tag{3.35}$$

subject to:

$$y_i \left(\mathbf{w}^\star \cdot \Phi(\mathbf{x}_i) + b^\star \right) \geq 1, \quad i = 1, \ldots, m. \tag{3.36}$$

where $(\mathbf{w}^\star, b^\star)$ denotes the solution. As we have seen, maximizing the margin subject to these constraints gives the following dual problem:

$$\omega(K) = \max_{\alpha} \left\{ \sum_{i=1}^{m} \alpha_i - \frac{1}{2} \sum_{i,j=1}^{m} \alpha_i \alpha_j y_i y_j K(x_i, x_j) : \sum_{i=1}^{m} \alpha_i y_i = 0, \alpha_i \geq 0 \right\} \tag{3.37}$$

If the dataset is linearly separable in feature space, this maximizing the margin criterion for kernel learning can be represented as the following min-max problem

$$\min_{K \in \mathcal{K}} \omega(K) = \min_{K \in \mathcal{K}} \max_{\alpha} \left\{ \sum_{i=1}^{m} \alpha_i - \frac{1}{2} \sum_{i,j=1}^{m} \alpha_i \alpha_j y_i y_j K(x_i, x_j) \right\} \tag{3.38}$$

subject to:

$$\sum_{i=1}^{m} \alpha_i y_i = 0 \qquad \alpha_i \geq 0 \tag{3.39}$$

For a non-linearly separable dataset, we can consider SVM binary classification with an L_1-norm soft margin, for example. In this case, we would consider minimizing the following dual problem to learn the kernel:

$$\min_{K \in \mathcal{K}} \omega(K) = \min_{K \in \mathcal{K}} \max_{\alpha} \left\{ \sum_{i=1}^{m} \alpha_i - \frac{1}{2} \sum_{i,j=1}^{m} \alpha_i \alpha_j y_i y_j K(x_i, x_j) \right\} \tag{3.40}$$

subject to:

$$\sum_{i=1}^{m} \alpha_i y_i = 0 \qquad 0 \leq \alpha_i \leq C \tag{3.41}$$

If $\mathcal{K}$ is the linear combination of kernel matrices, given by equation (3.34), this maximum margin approach to kernel combination learning reduces to:

$$\min_{\lambda} \max_{\alpha} \left\{ \sum_{i=1}^{m} \alpha_i - \frac{1}{2} \sum_{i,j=1}^{m} \alpha_i \alpha_j y_i y_j \left[\sum_{\ell=1}^{p} \lambda_\ell K_\ell(x_i, x_j) \right] \right\} \tag{3.42}$$

subject to

$$\sum_{i=1}^{m} \alpha_i y_i = 0 \qquad 0 \leq \alpha_i \leq C \tag{3.43}$$

$$\sum_{\ell=1}^{p} \lambda_\ell = 1 \qquad \lambda_\ell \geq 0. \tag{3.44}$$

Further reading: It is worth noting that Lanckriet et al. [2004a] considered a general formulation where $\{K_\ell\}$ does not need to be positive semi-definite nor do the kernel combination coefficients $\{\lambda_\ell\}$ have to be non-negative. However, the performance of this more general formulation is no better than the specific formulation in (3.34) with $\lambda_\ell \geq 0$. This more general formulation leads to a *semi-definite programming* (SDP) problem, which does not scale well to large datasets. For simplicity, we restrict our discussion to the MKL formulation in (3.34).

3.8   ALGORITHMIC APPROACHES TO MULTIPLE KERNEL LEARNING

Having stated the objective function we wish to optimize, we now come to the issue of an effective algorithmic procedure to do this. Let us denote the set of λ_ℓ satisfying the given constraints by:

$$\triangle = \left\{ \lambda = (\lambda_1, \ldots, \lambda_p) : \sum_{\ell=1}^{p} \lambda_\ell = 1, \ \lambda_\ell \geq 0, \ \forall \ell \right\} \tag{3.45}$$

Similarly the set of α_i which satisfy their given constraints are:

$$\mathcal{Q} = \left\{ \alpha = (\alpha_1, \ldots, \alpha_m) : \sum_{i=1}^{m} \alpha_i y_i = 0, \ 0 \leq \alpha_i \leq C, \ \forall i \right\}. \tag{3.46}$$

Furthermore, let us denote the objective function in the min-max formulation (3.42) by:

$$\mathcal{L}(\alpha, \lambda) = \sum_{i=1}^{m} \alpha_i - \frac{1}{2} \sum_{i,j=1}^{m} \alpha_i \alpha_j y_i y_j \left[\sum_{\ell=1}^{p} \lambda_\ell K_\ell(x_i, x_j) \right]. \tag{3.47}$$

Then, the dual objective optimization (3.34) can be simply represented by:

$$\min_{\lambda \in \triangle} \max_{\alpha \in \mathcal{Q}} \mathcal{L}(\alpha, \lambda). \tag{3.48}$$

Quadratically Constrained Linear Programming (QCLP): It is easy to see that $\mathcal{L}(\alpha, \lambda)$ is *concave* with respect to α and *convex* with respect to λ. The *min-max theorem* states that:

$$\min_{\lambda \in \triangle} \max_{\alpha \in \mathcal{Q}} \mathcal{L}(\alpha, \lambda) = \max_{\alpha \in \mathcal{Q}} \min_{\lambda \in \triangle} \mathcal{L}(\alpha, \lambda). \tag{3.49}$$

Let us consider the minimization $\min_{\lambda \in \triangle} \mathcal{L}(\alpha, \lambda)$ for fixed $\alpha \in \mathcal{Q}$. It is a linear programming problem with respective to λ. Let us write $S_\ell(\alpha) = \frac{1}{2} \sum_{i,j} \alpha_i \alpha_j y_i y_j K_\ell(\mathbf{x}_i, \mathbf{x}_j)$ and let us define u by $u = \sum_\ell S_\ell(\alpha) \lambda_\ell$, then the condition $\sum_\ell \lambda_\ell = 1$ means that λ_ℓ is at most equal to 1 and thus that $S_\ell(\alpha) \leq u$. Thus we infer that this linear programming problem is equivalent to:

$$\max_t \sum_{i=1}^{m} \alpha_i - u \tag{3.50}$$

subject to:

$$\sum_{i,j=1}^{m} \alpha_i \alpha_j y_i y_j K_\ell(x_i, x_j) \leq 2u \ \forall \ell. \tag{3.51}$$

supplemented by the conditions:

$$\sum_{i=1}^{m} \alpha_i y_i = 0 \qquad\qquad 0 \le \alpha_i \le C. \qquad\qquad (3.52)$$

if an L_1 soft margin is used. With respect to both variables, the above objective function is *linear* and the constraints $\sum_{i,j=1}^{m} \alpha_i \alpha_j y_i y_j K_\ell(x_i, x_j) \le 2u$ are *quadratic*. This type of problem is usually referred to as *quadratically constrained linear programming* (QCLP).

Further reading: this convex QCLP problem is a special instance of a second-order cone programming problem (SOCP), which is also a special form of semi-definite programming (SDP) (Vandenberghe and Boyd [1996]). SOCPs can be efficiently solved with general purpose software such as SeDumi (Sturm [1999]) or Mosek (Anderson and Anderson [2000]). These codes employ interior-point methods (Nesterov and Nemirovsky [1994]), which has a worst-case complexity $\mathcal{O}(pm^3)$. The use of such algorithms will only be feasible for small problems with few data points and kernels. Bach et al. [2004] suggested an algorithm based on sequential minimization optimization (SMO). While the MKL problem is convex, it is also non-smooth, making the direct application of simple local descent algorithms such as SMO infeasible. Bach et al. [2004] therefore considered a smoothed version of the problem to which SMO can be applied.

Semi-infinite Linear Programming (SILP): We can also reformulate the binary classification MKL problem as follows. Let $t = \max_{\alpha \in \mathcal{Q}} \mathcal{L}(\alpha, \lambda)$. Hence, the problem (3.48) is equivalent to

$$\min_{t} t \qquad\qquad (3.53)$$

subject to:

$$t = \max_{\alpha \in \mathcal{Q}} \mathcal{L}(\alpha, \lambda) \qquad\qquad (3.54)$$

Since we are minimizing over t, this problem can be restated as:

$$\min_{t} t \qquad\qquad (3.55)$$

subject to:

$$\mathcal{L}(\alpha, \lambda) \le t, \quad \forall \alpha \in \mathcal{Q}. \qquad\qquad (3.56)$$

Let $S_0(\alpha) = \sum_i \alpha_i$ and $S_\ell(\alpha) = \frac{1}{2} \sum_{i,j} \alpha_i \alpha_j y_i y_j K_\ell(x_i, x_j)$, then the complete problem can be stated as:

$$\min_{\lambda, t} t \qquad\qquad (3.57)$$

subject to:

$$\sum_\ell \lambda_\ell = 1, \quad 0 \le \lambda \le 1 \tag{3.58}$$

$$S_0(\alpha) - \sum_\ell \lambda_\ell S_\ell(\alpha) \le t, \forall \alpha \in \mathcal{Q}. \tag{3.59}$$

We refer to the above formulation as a *semi-infinite linear programming (SILP) problem* since the objective function is linear, and there are infinite number of linear constraints with respect to λ and t, which is indexed by $\alpha \in \mathcal{Q}$. A SILP problem can be solved by an iterative algorithm called *column generation* (or exchange methods), which is guaranteed to converge to a global optimum. The basic idea is to compute the optimum (λ, t) by linear programming for a restricted subset of constraints and update the constraint subset based on the obtained suboptimal (λ, t). In particular, given a set of restricted constraints $\{\alpha_q : q = 1 \ldots, Q\}$, first we find the intermediate solution (λ, t) by the following linear programming optimization with Q linear constraints

$$\min_{t, \lambda} t \tag{3.60}$$

subject to:

$$\sum_\ell \lambda_\ell = 1, \quad 0 \le \lambda \le 1 \tag{3.61}$$

$$S_0(\alpha_q) - \sum_\ell \lambda_\ell S_\ell(\alpha_q) \le t, \forall q = 1, \ldots, Q. \tag{3.62}$$

This problem is often called the *restricted master problem*. For the given intermediate solution (λ, t), we find the next constraint with the maximum violation, i.e., $\max_\alpha S_0(\alpha) - \sum_\ell \lambda_\ell S_\ell(\alpha)$, which is a standard SVM optimization problem with kernel matrix $\left(\sum_\ell \lambda_\ell K_\ell(x_i, x_j) \right)_{i,j=1}^m$. If the optimal $\alpha^\star$ satisfies $S_0(\alpha^\star) - \sum_\ell \lambda_\ell S_\ell(\alpha^\star) \le t$, then the current intermediate solution (λ, t) is optimal for the optimization (3.57-3.59). Otherwise, $\alpha^\star$ should be added to the restriction set. We repeat the above iteration until convergence, which is guaranteed to be globally optimal. The convergence criterion for the SILP is usually chosen as

$$\left| 1 - \frac{\sum_\ell \lambda_\ell^{(t-1)} S_\ell(\alpha^{(t)}) - S_0(\alpha^{(t)})}{\gamma^{(t-1)}} \right| \le \epsilon. \tag{3.63}$$

Hence, if we have an efficient LP solver and a SVM algorithm for a single kernel, much larger datasets can be handled than the QCLP approach. Using this general algorithm, it was possible to solve MKL problems with up to $30,000$ datapoints and 20 kernels within a reasonable time.

Further reading: MKL based on semi-infinite linear programming (Hettich and Kortanek [1993]) was initially considered by Sonnenburg et al. [2006], who also investigated its extension to handling regression and one-class classification.

First-order Gradient Methods: The SILP approach can handle large scale datasets since it only needs an efficient LP solver and a SVM solver with a single kernel. However, SILP uses the *cutting plane approach* which could take a long time to converge, especially for a large number of kernels. Another alternative approach is to use *first-order gradient methods*. Let:

$$J(\lambda) = \max_{\alpha \in \mathcal{Q}} \mathcal{L}(\alpha, \lambda). \qquad (3.64)$$

We assume that each matrix $(K_\ell(x_i, x_j))_{ij}$ is positive definite, with all eigenvalues greater than zero. This can be guaranteed by enforcing a small *ridge* to the diagonal of the matrix using:

$$K_\ell(x_i, x_i) \longleftarrow K_\ell(x_i, x_i) + \delta, \quad \forall i, \ell, \qquad (3.65)$$

where $\delta > 0$ is the small ridge constant. This implies that, for any admissible value of λ, $\mathcal{L}(\cdot, \lambda)$ is strictly concave. In turn, this property of strict concavity ensures that $J(\lambda)$ is convex and differentiable.
Further reading: The existence and computation of derivatives of optimal value functions such as $J(\lambda)$ have been discussed in the literature (Bonnans and Shapiro [1998], Chapelle et al. [2002], Rakotomamonjy et al. [2008], Ying et al. [2009a]). Rakotomamonjy et al. [2008] pointed out that the projected gradient method (e.g., Vandenberghe and Boyd [1996]) could take a long time to converge. Instead, they proposed to employ the *reduced gradient method*, an approach they named *SimpleMKL*. Various experiments were conducted to show that SimpleMKL is more efficient than the SILP approach. Another MKL algorithm have been developed by Xu et al. [2008], using an extended level method. Indeed, since $\mathcal{L}(\cdot, \lambda)$ is strongly concave for any $\lambda \in \Delta$, using a similar argument in Lemaréchal and Sagastizabal [1997], Ying et al. [2009a], we can easily prove that $J(\cdot)$ is differentiable and has a Lipschitz gradient. Then again, it also be possible to apply Nesterov's smooth optimization method (Nesterov [2003]), which has a convergence rate $\mathcal{O}(1/k^2)$, where k is the iteration number. Two further efficient methods have been proposed by Ying et al. [2009b], called *MKLdiv-conv* (based on a convex maximum determinant problem) and *MKLdiv-dc* (based on a difference of convex formulation).

3.9   CASE STUDY 4: PROTEIN FOLD PREDICTION

Proteins are the functional molecular components inside cells. Messenger RNA is *transcribed* from a genetic sequence, which *codes* for that protein. This RNA is then *translated* into protein. An important final step in this process is *folding* in which the protein forms its final three-dimensional structure. Understanding the three-dimensional structure of a protein can give insight into its function. If the protein is a drug target, knowledge of its structure is important in the design of small molecular inhibitors which would bind to, and disable, the protein. Advances in gene sequencing technologies have resulted in a large increase in the number of identified sequences which code for proteins.

However, there has been a much slower increase in the number of known three-dimensional protein structures.

This motivates the problem of using machine learning methods to predict the structure of a protein from sequence and other data. In this Case Study, we will only consider a sub-problem of structure prediction in which the predicted label is over a set of *fold classes*. The fold classes are a set of structural components, common across proteins, which give rise to the overall three-dimensional structure. In this study, there were 27 fold classes with 313 proteins used for training and 385 for testing. There are a number of observational features relevant to predicting fold class, and in this study, we used 12 different informative data-types, or feature spaces. These included the RNA sequence and various physical measurements such as hydrophobicity, polarity and van der Waals volume. Viewed as a machine learning task, there are multiple data-types, thus we consider multiple kernel learning. Prediction is over 27 classes. Our discussion of MKL algorithms in the previous Sections was geared to binary classification. Thus we use a different MKL algorithms which can handle multi-class classification (see Further reading).

In Figure 3.1, we illustrate the performance of a MKL algorithm (*MKLdiv-dc*) on this dataset. In Figure 3.1 (left), the vertical bars indicate the test set accuracy based on using one type of data only: for example, H is Hydrophobicity, P is Polarity, V is van der Waals volume (V). The horizontal bar indicates the performance of the MKL algorithm with all data-types included. Plainly, we get an improvement in performance if we use all available data over just using the single most informative data-source.

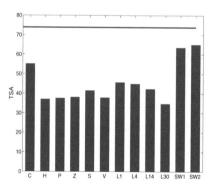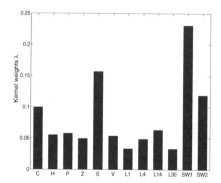

Figure 3.1:   Performance of a MKL algorithm (*MKLdiv-dc*) on a protein fold prediction dataset. There are 27 classes and 12 types of input data. *Left*: test set accuracy (TSA, as a %) based on individual datatypes (vertical bars) and using MKL (horizontal bar). *Right*: the kernel coefficients λ_ℓ, which indicate the relative significance of individual types of data.

Figure 3.1 (right) gives the values of the kernel coefficients λ_ℓ, based on using a linear combination (3.34). The relative height of the peaks indicates the relative significance of different types of input data. This algorithm indicates that all 12 types of data are relevant, though some types of data are more informative than others (the most informative $SW1$ and $SW2$ are based on sequence alignments).

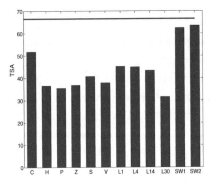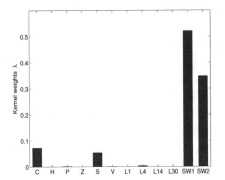

Figure 3.2:   Performance of a MKL algorithm (*SimpleMKL*) on a protein fold prediction dataset. *Left*: test set accuracy (TSA, as a %). *Right*: the kernel coefficients λ_ℓ. This algorithm is less accurate than *MKLdiv-dc*, but it entirely eliminates certain datatypes (right figure). Thus lower accuracy is achieved based on the use of fewer types of data.

A datatype can be weakly informative for class discrimination. In addition, some types of data may be implicit in another type of data. Thus this data source could be eliminated with little or no cost to performance. Different MKL algorithms differ in the extent to which they remove apparently redundant data sources. In Figure 3.2, we use the same protein fold dataset with a different algorithm: *SimpleMKL*. In this case, the overall test set accuracy is lower, but certain types of data have been entirely eliminated (Figure 3.2 (right)). This could be a beneficial trade-off in that we do not need to conduct certain measurements at the expense of a small loss in accuracy.

Further reading: MKL methods have been successfully demonstrated on the integration of multiple heterogeneous data sources to enhance inference with biological datasets (Damoulas and Girolami [2008], Kloft et al. [2009], Lanckriet et al. [2004a,b], Ying et al. [2009b]).

APPENDIX A

Appendix

A.1 INTRODUCTION TO OPTIMIZATION THEORY

Support Vector Machines and other kernel-based methods make extensive use of optimization theory. In this Appendix, we give an introductory description of some elements of optimization theory most relevant to this subject. One of the principal applications of optimization theory is the maximization or minimization of a function of one or more variables (the *objective function*). This optimization task may be subject to possible *constraints*. An optimization problem can be classified according to the type of objective function and constraints considered. With *linear programming* (LP), the objective function and any constraints are linear functions of the variables, and these variables are real-valued. With *nonlinear programming*, the objective function and/or constraints are nonlinear functions of the variables. For *integer programming* (IP), the variables are fixed to take integer values. *Dynamic programming* is a different type of optimization problem in which we have a number of subtasks (e.g., good or poor moves in a game) and an eventual goal (win or loose the game). Dynamic programming is relevant to the task of constructing sequence kernels mentioned in Section 3.4.

In Section 2.2, we mentioned a linear programming approach to constructing a classifier. The general form of a linear programming problem is:

$$\text{max or min} \left[\sum_{i=1}^{n} c_i x_i \right] \tag{A.1}$$

subject to constraints:

$$\sum_{j=1}^{n} a_{ij} x_i \left. \begin{array}{c} \geq \\ = \\ \leq \end{array} \right\} b_i \qquad i = 1, 2, \ldots, m \tag{A.2}$$

and non-negativity constraints $x_i \geq 0$, $i = 1, 2 \ldots n$. We can introduce a *slack variable* to make an inequality such as $x_1 \leq b_1$ into an equality $x_1 + s_1 = b_1$ and similarly a *surplus variables* to make $x_1 \geq b_1$ into $x_1 - s_1 = b_1$. Thus, expressed in matrix and vector form, the general statement of a LP problem is:

$$\max_{\mathbf{x}} \left[\mathbf{c}^T \mathbf{x} \right] \qquad \text{or} \qquad \min_{\mathbf{x}} \left[\mathbf{c}^T \mathbf{x} \right] \tag{A.3}$$

subject to a set of equality constraints, given by $\mathbf{Ax} = \mathbf{b}, \mathbf{x} \geq \mathbf{0}$. The linear equalities in $\mathbf{Ax} = \mathbf{b}$ can be viewed as a set of bounding hyperplanes defining a *feasible region* within which the optimal solution must lie. This is pictured in Figure A.1 (left), where the feasible region, is depicted as a closed region, though it can be open in general. The objective function is $S = \sum_{i=1}^{n} c_i x_i$ and different values of S define a series of lines of constant gradient which cross the feasible region. An immediate observation is that the minimal or maximal value of the objective function must either be a unique solution, corresponding to an edge point of the feasible region, or multiple optimal solutions with the same value of the objective function corresponding to all points along an edge of the feasible region. The feasible region is said to constitute a *convex* set of points since, if we draw a line connecting any two points in the region, the line will lie wholly within or along a edge of the feasible region. This is not true for Figure A.1 (right) where a line connecting points A and B would pass outside the region: this is a *non-convex* set of points. This property of convexity is true for all linear programming problems. However, for nonlinear programming with a non-convex feasible region, it is possible to get caught in sub-optimal local maxima depending on the starting point of a iterative optimization algorithm.

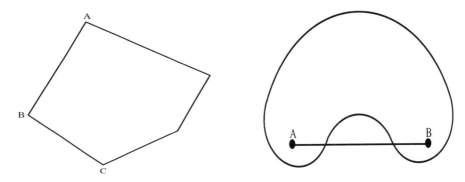

Figure A.1:   The feasible region of an LP problem is convex (*left*): if we draw a line between two points it remains within the feasible region. For a non-convex space (*right*), we can draw a line between two points (e.g., A and B), which will pass outside the feasible region.

We can formally define *convexity* as follows: if $\mathbf{x}_1$ and $\mathbf{x}_2$ are two points in the region, then so is every point $\lambda \mathbf{x}_2 + (1 - \lambda)\mathbf{x}_1$ where $0 < \lambda < 1$. A function $f(\mathbf{x})$ is then said to be *convex* over a convex domain X if for any two points $\mathbf{x}_1, \mathbf{x}_2 \in X$:

$$f[\mathbf{x}] = f[\lambda \mathbf{x}_2 + (1 - \lambda)\mathbf{x}_1] \leq \lambda f(\mathbf{x}_2) + (1 - \lambda)f(\mathbf{x}_1) \qquad (A.4)$$

for $0 \leq \lambda \leq 1$. Thus, for a function of one variable, this means that $f[x]$ lies below the chord joining any two points on its graph: an example of a convex function would be $f(x) = x^2$. For a *concave* function, we reverse the inequality sign in (A.4): an example of a concave function would be $f(x) = -x^2$. The Support Vector Machine has a quadratic objective function, which therefore satisfies the property of concavity (strictly, the constraint $\sum_{i=1}^{m} y_i \alpha_i = 0$ could give a degeneracy of solutions with the same value of the objective function in rare instances). This feature of a unique optimal solution is one of the principal attractions of SVM learning.

Apart from linear and quadratic programming approaches to kernel-based learning, we also mentioned other optimization topics, principally in the context of multiple kernel learning. *Semi-infinite linear programming* (SILP) involves a finite number of variables with an infinite number of constraints or an infinite number of variables with a finite number of constraints. *Quadratically constrained linear programming* (QCLP) involves a linear objective function and quadratic constraints. Convex QCLP problems are a special instance of *second-order cone programming* (SOCP) problems which are, in turn, a special instance of *semi-definite programming* (SDP) problems. SOCP problems can be efficiently solved using general-purpose software packages such as *SeDumi* or *Mosek*.

Further Reading: there are many textbooks covering linear and nonlinear programming (Bertsekas [2004], Bazarra et al. [2006], Luenberger [2003]). Semi-infinite linear programming is described in Hettich and Kortanek [1993]. Vandenberghe and Boyd [1996] gives a survey of conic optimization.

A.2   DUALITY

The concept of duality is important in many areas of mathematics. We can consider *geometric duality*. Thus, in 2 dimensions, two non-parallel lines uniquely determine a point at their intersection. Similarly two non-identical points uniquely determine a line passing through the 2 points. This concept generalizes to many dimensions and can give insights into SVM learning. Thus, when we introduced SVMs for binary classification, we pictured the classifier as a hyperplane in the data space, with the data represented as points, as illustrated in Figure 1.1 (left) for a separable dataset. Potentially there is an infinite family of hyperplanes which would label the two classes of data correctly. In the dual space (which we call *version space*), hyperplanes become points and points become hyperplanes. This is illustrated in Figure A.2 (right) where the space is pictured as closed: all points within this space correspond to separating hyperplanes in the original data space, which we will now call *instance space*. The bounding hyperplanes in version space correspond to datapoints in instance space.

In version space, the solution determined by an SVM corresponds to the largest inscribed hypersphere bounded by the hyperplanes within version space. This inscribed hypersphere will touch certain walls within version space, and these correspond to the support vectors back in instance space. The SVM solution looks effective, but it does not look like the optimal solution. Specifically, any point (hypothesis) in version space which is away from the center and close to several of the bounding hyperplanes looks like a biased solution. Back in instance space, it would correspond to a solution which is close to one class of data and away from the other class (depicted by the dashed hyperplane in Figure A.2 (*left*)).

From the viewpoint of version space: what is the best type of classifier? Intuitively, it appears to be the center of mass of version space, which appears to be the least biased solution. Formally, we can define such a solution as follows. Imagine a hyperplane which cuts version space into two equal halves.

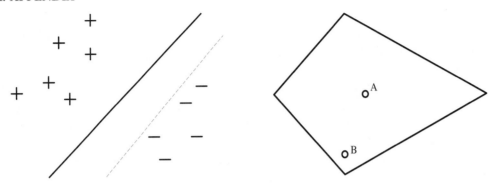

Figure A.2: *Left*: datapoints separated by directed hyperplanes: datapoints on one side are labelled +1 and the other −1. *Right*: in a dual representation points become hyperplanes and hyperplanes become points: all points within the enclosed region are separating hyperplanes in instance space (left figure). The center of mass (A) is an unbiased solution but a choice such as B corresponds to a biased selection corresponding to a poor separatrix in instance space (dashed line, left figure).

This appears to be an unbiased construction. If there was a point in version space corresponding to the exact solution, then it could lie in either half with equal probability. Unfortunately, there is an infinite family of such bisecting hyperplanes. They do not meet at a point but in a small region, and the center of this region is the *Bayes point*. The Bayes point is approximately at the center of mass of this space and thus learning machines based on finding the Bayes point or center of mass are potentially more powerful than SVMs (see Section 2.2).

In optimization theory, duality can be formally defined for linear programming, for example. We call the following pair of problems a *primal-dual pair*:

Primal Problem	**Dual Problem**
$\min_{\mathbf{x}} \left[S_x = \mathbf{c}^T \mathbf{x} \right]$	$\max_{\mathbf{y}} \left[S_y = \mathbf{b}^T \mathbf{y} \right]$
subject to $\mathbf{Ax} \geq \mathbf{b}$	subject to $\mathbf{A}^T \mathbf{y} \leq \mathbf{c}$
$\mathbf{x} \geq \mathbf{0}$	$\mathbf{y} \geq \mathbf{0}$

Thus for every primal linear programming problem, there is a dual formulation. If solved, this dual problem can give the solution to the original primal problem. Primal and dual are closely related and have the important property that the optimal values of both objective functions are equal at the solution. For a Support Vector Machine, the primal formulation involved optimization over the weight vector $\mathbf{w}$ while the dual involved optimization over α_i. Interestingly, the dimensionality of $\mathbf{w}$ is the number of features whereas α_i is indexed by the number of samples. This is true for LP primal-dual pair stated above: $\mathbf{x}$ and $\mathbf{y}$ do not have the same dimensionality, in general. This can be important in practical applications. Thus, for the microarray dataset for Wilm's tumor mentioned in Section 1.7, the number of features is 17,790, but the number of samples is 27. Thus, for the dual problem, the optimization process is less computationally intensive since it is performed over fewer parameters.

A.3   CONSTRAINED OPTIMIZATION

When we introduced the Support Vector Machine in Chapter 1, we noted that the learning task involved optimization of an objective function (1.7) subject to constraints (1.8). In the context of optimization theory, this required the introduction of Lagrange multipliers α_i to handle the constraints. In this Section, we briefly review constrained optimization to provide some further mathematical background. For purposes of illustration, let us consider a simple two-variable optimization problem with objective function $z = f(x, y)$ and with the component variables x and y constrained to satisfy the equality constraint $g(x, y) = 0$

This equation implicitly defines a relation between x and y, which we could write $y = h(x)$, and thus:

$$z = f(x, y) = f(x, h(x)) \tag{A.5}$$

which is now a function of one independent variable x. At an optimum of z, we have:

$$\frac{dz}{dx} = \frac{\partial f}{\partial x} + \frac{\partial f}{\partial y}\frac{dy}{dx} = 0 \tag{A.6}$$

Furthermore, from $dg(x, y) = (\partial g/\partial x)dx + (\partial g/\partial y)dy = 0$, we deduce:

$$\frac{dy}{dx} = -\left[\frac{\partial g}{\partial y}\right]^{-1}\frac{\partial g}{\partial x} \tag{A.7}$$

Thus:

$$\frac{\partial f}{\partial x} - \frac{\partial f}{\partial y}\left[\frac{\partial g}{\partial y}\right]^{-1}\frac{\partial g}{\partial x} = 0 \tag{A.8}$$

If we define:

$$\lambda = -\left[\frac{\partial g}{\partial y}(x, y)\right]^{-1}\left[\frac{\partial f}{\partial y}(x, y)\right] \tag{A.9}$$

then, at the optimum, we have:

$$\frac{\partial f}{\partial x}(x, y) + \lambda\frac{\partial g}{\partial x}(x, y) = 0 \tag{A.10}$$

A similar argument with x and y interchanged gives:

$$\frac{\partial f}{\partial y}(x, y) + \lambda\frac{\partial g}{\partial y}(x, y) = 0 \tag{A.11}$$

and, of course, we have $g(x, y) = 0$. These three conditions can be generated from a *Lagrange function*:

$$F(x, y, \lambda) = f(x, y) + \lambda g(x, y) \tag{A.12}$$

as the three derivatives with respect to x, y and λ, respectively. λ is a *Lagrange multiplier* and the objective function has been restated as the sum of the original objective and λ multiplying the equality constraint. This generalizes to n variables and m equality constraints with corresponding Lagrange function:

$$F(\mathbf{x}, \lambda) = f(\mathbf{x}) + \sum_{i=1}^{m} \lambda_i g_i(\mathbf{x}) \tag{A.13}$$

We can also readily incorporate inequality constraints. Since a constraint $c_i(\mathbf{x}) \geq b_i$ can be viewed as $-c_i(\mathbf{x}) \leq -b_i$, we can generically consider all inequalities as of the form:

$$g_i(\mathbf{x}) \leq b_i \tag{A.14}$$

These inequalities can be transformed into equality constraints by addition of non-negative *slack variables* u_i^2 (squared to ensure positivity):

$$g_i(\mathbf{x}) + u_i^2 - b_i = 0 \tag{A.15}$$

and so the problem is to optimize $f(x)$ subject to the equality constraints $g_i(\mathbf{x}) + u_i^2 - b_i = 0$ with Lagrange function:

$$F(\mathbf{x}, \lambda, \mathbf{u}) = f(\mathbf{x}) + \sum_{i=1}^{m} \lambda_i \left[g_i(\mathbf{x}) + u_i^2 - b_i \right] \tag{A.16}$$

Necessary conditions to be satisfied at a stationary point are then:

$$\frac{\partial F}{\partial x_j} = 0 = \frac{\partial f}{\partial x_j} + \sum_{i=1}^{m} \lambda_i \frac{\partial g_i}{\partial x_j}; \qquad j = 1, 2, \ldots, n \tag{A.17}$$

$$\frac{\partial F}{\partial \lambda_i} = 0 = g_i(\mathbf{x}) + u_i^2 - b_i; \qquad i = 1, 2, \ldots, m \tag{A.18}$$

$$\frac{\partial F}{\partial u_i} = 0 = 2\lambda_i u_i \qquad i = 1, 2, \ldots, m \tag{A.19}$$

The last condition, when multiplied by $u_i/2$, gives:

$$\lambda_i u_i^2 = 0 \tag{A.20}$$

i.e.,:

$$\lambda_i \left[b_i - g_i(\mathbf{x}) \right] = 0 \tag{A.21}$$

This last equation states that either λ_i or $(b_i - g_i(\mathbf{x}^\star))$ is zero, where $\mathbf{x}^\star$ denotes the value of $\mathbf{x}$ at this optimum. If the Lagrange multiplier λ_i is not zero, then $g_i(\mathbf{x}^\star) = b_i$, and the constraint is said to be *active*. On the other hand, if $g_i(\mathbf{x}^\star) < b$, then the corresponding Lagrange multiplier λ_i must be zero. For a Support Vector Machine, when the Lagrange multiplier α_i is non-zero, then $y_i [\mathbf{w} \cdot \mathbf{z} + b] = 1$ corresponding to i being a support vector. When $\alpha_i = 0$ then $y_i [\mathbf{w} \cdot \mathbf{z} + b] > 1$ and i is a non-support vector and this datapoint does not make a contribution to the argument inside the decision function.

The Lagrange multipliers are constrained in sign as a consequence of the following argument. Since $g_k(\mathbf{x}) + u_k^2 = b_k$:

$$\frac{\partial g_k}{\partial b_i} = \begin{cases} 0 & \text{if } i \neq k \\ 1 & \text{if } i = k \end{cases} \tag{A.22}$$

Thus if we consider the optimum of the Lagrange function with respect to b_i:

$$\frac{\partial F}{\partial b_i} = \frac{\partial f}{\partial b_i} + \sum_{k=1}^{m} \lambda_k^\star \frac{\partial g_k}{\partial b_i} = \frac{\partial f}{\partial b_i} + \lambda_i^\star = 0 \tag{A.23}$$

where $\lambda_i^\star$ denotes the value of λ_i at this optimum. Thus:

$$\frac{\partial f}{\partial b_i} = -\lambda_i^\star \tag{A.24}$$

If b_i is increased, the constraint region is enlarged so the minimum of $f(\mathbf{x})$ could remain the same or it could decrease in value. We thus deduce that:

$$\frac{\partial f}{\partial b_i} \leq 0 \tag{A.25}$$

and hence that $\lambda_i^\star \geq 0$. Thus, for inequality constraints, at a minimum of $f(\mathbf{x})$ the following conditions are satisfied:

$$
\begin{aligned}
\frac{\partial f}{\partial x_j} + \sum_{i=1}^{m} \lambda_i \frac{\partial g_i}{\partial x_j} &= 0 & j &= 1, 2, \ldots, n. \\
g_i(\mathbf{x}) &\leq b_i & i &= 1, 2, \ldots, m \\
\lambda_i [g_i(\mathbf{x}) - b_i] &= 0 & i &= 1, 2, \ldots, m \\
\lambda_i &\geq 0 & i &= 1, 2, \ldots, m
\end{aligned}
\tag{A.26}
$$

This is an example of the *Karush-Kuhn-Tucker (KKT) conditions*: the complete set of necessary conditions to be satisfied at an optimum. For maximization, this argument would lead to $\lambda_i \leq 0$.

However, for our statement of the SVM learning task in (1.4), we have *subtracted* the second term which incorporates the Lagrange multipliers and hence the requirement is $\alpha_i \geq 0$.

Given the primal formulation of the SVM classifier in (1.4) :

$$L = \frac{1}{2}(\mathbf{w} \cdot \mathbf{w}) - \sum_{i=1}^{m} \alpha_i \left(y_i (\mathbf{w} \cdot \mathbf{x} + b) - 1 \right) \qquad \text{(A.27)}$$

The KKT conditions are therefore:

$$
\begin{aligned}
\frac{\partial L}{\partial \mathbf{w}} &= \mathbf{w} - \sum_{i=1}^{m} \alpha_i\, y_i \mathbf{x}_i = 0 \\
\frac{\partial L}{\partial b} &= -\sum_{i=1}^{m} \alpha_i\, y_i = 0 \\
y_i (\mathbf{w} \cdot \mathbf{x}_i + b) &\geq 1 \\
\alpha_i \left(y_i (\mathbf{w} \cdot \mathbf{x}_i + b) - 1 \right) &= 0 \qquad \text{(A.28)} \\
\alpha_i &\geq 0
\end{aligned}
$$

For regression and the primal formulation in (2.41,2.42), the associated Lagrange function is:

$$
\begin{aligned}
L &= \mathbf{w} \cdot \mathbf{w} + C \sum_{i=1}^{m} \left(\xi_i + \widehat{\xi}_i \right) - \sum_{i=1}^{m} \alpha_i \left(\epsilon + \xi_i - y_i + \mathbf{w} \cdot \mathbf{x}_i + b \right) \\
&\quad - \sum_{i=1}^{m} \widehat{\alpha}_i \left(\epsilon + \widehat{\xi}_i + y_i - \mathbf{w} \cdot \mathbf{x}_i - b \right) - \sum_{i=1}^{m} \beta_i \xi_i - \sum_{i=1}^{m} \widehat{\beta}_i \widehat{\xi}_i
\end{aligned}
$$

since there are 4 constraints:

$$
\begin{aligned}
y_i - \mathbf{w} \cdot \mathbf{x}_i - b &\leq \epsilon + \xi_i & \text{(A.29)} \\
(\mathbf{w} \cdot \mathbf{x}_i + b) - y_i &\leq \epsilon + \widehat{\xi}_i & \text{(A.30)} \\
\xi_i, \widehat{\xi}_i &\geq 0 & \text{(A.31)}
\end{aligned}
$$

The KKT conditions are:

$$\frac{\partial L}{\partial \mathbf{w}} = \mathbf{w} - \sum_{i=1}^{m} \alpha_i \mathbf{x}_i + \sum_{i=1}^{m} \widehat{\alpha}_i \mathbf{x}_i = 0$$

$$\frac{\partial L}{\partial b} = -\sum_{i=1}^{m} \alpha_i + \sum_{i=1}^{m} \widehat{\alpha}_i = 0$$

$$\frac{\partial L}{\partial \xi_i} = C - \alpha_i - \beta_i = 0$$

$$\frac{\partial L}{\partial \widehat{\xi}_i} = C - \widehat{\alpha}_i - \widehat{\beta}_i = 0$$

$$y_i - \mathbf{w} \cdot \mathbf{x}_i - b \leq \epsilon + \xi_i$$

$$(\mathbf{w} \cdot \mathbf{x}_i + b) - y_i \leq \epsilon + \widehat{\xi}_i$$

$$\xi_i \geq 0 \qquad \widehat{\xi}_i \geq 0$$

$$\alpha_i \left(\epsilon + \xi_i - y_i + \mathbf{w} \cdot \mathbf{x}_i + b \right) = 0$$

$$\widehat{\alpha}_i \left(\epsilon + \widehat{\xi}_i + y_i - \mathbf{w} \cdot \mathbf{x}_i - b \right) = 0$$

$$\beta_i \xi_i = 0 \qquad \widehat{\beta}_i \widehat{\xi} = 0$$

$$\alpha_i \geq 0 \qquad \widehat{\alpha}_i \geq 0 \qquad \beta_i \geq 0 \qquad \widehat{\beta}_i \geq 0$$

from these conditions, we also deduce that:

$$(C - \alpha_i)\xi_i = 0$$

$$(C - \widehat{\alpha}_i)\widehat{\xi}_i = 0$$

which are used in (2.49).

Bibliography

E.L. Allwein, R.E. Schapire, and Y. Singer. Reducing multiclass to binary: a unifying approach for margin classifiers. *Journal of Machine Learning Research*, 1:133–141, 2000. http://www.jmlr.org 8

E.D. Anderson and A.D. Anderson. The MOSEK interior point optimizer for linear programming: An implementation of the homogeneous algorithm. In T. Terlaky H. Frenk, C. Roos and S. Zhang, editors, *High Performance Optimization*, pages 197–232. Kluwer Academic Publishers, 2000. 60

F. Bach. Consistency of the group lasso and multiple kernel learning. *Journal of Machine Learning Research*, 9:1179–1225, 2008. 57

F. Bach, G.R.G. Lanckriet, and M.I. Jordan. Multiple kernel learning, conic duality and the SMO algorithm. In *Proceedings of the Twenty-first International Conference on Machine Learning (ICML)*, 2004. DOI: 10.1145/1015330.1015424 57, 60

G. Bakir, B. Taskar, T. Hofmann, B. Schölkopf, A. Smola, and S.V.N Viswanathan. *Predicting structured data*. MIT Press, 2007. 43

G. Baudat and F. Anouar. Generalised discriminant analysis using a kernel approach. *Neural Computation*, 12:2385–2404, 2000. DOI: 10.1162/089976600300014980 28

M.S. Bazarra, H.D. Sherali, and C.M. Shetty. *Nonlinear Programming, Theory and Algorithms*. Wiley, 2006. 67

K.P. Bennett. Combining support vector and mathematical programming methods for induction. In *Advances in Kernel Methods*, pages 307–326. MIT Press, 1999. 28

D.P. Bertsekas. *Nonlinear Programming*. Athena Scientific, 2004. 67

J.F. Bonnans and A. Shapiro. Optimization problems with perturbation: a guided tour. *SIAM Review*, 40:202–227, 1998. DOI: 10.1137/S0036144596302644 62

K.M. Borgwardt, C.S. Ong, S. Schönauer, and S. V. N. Vishwanathan. Protein function prediction via graph kernels. In *Proceedings of Intelligent Systems in Molecular Biology (ISMB)*. Detroit, USA, 2005. DOI: 10.1093/bioinformatics/bti1007 56

L. Bottou, O. Chapelle, D. DeCoste, and J. Weston. *Large-Scale Kernel Machines*. Neural Information Processing Series, The MIT Press, 2007. 15, 17

P.S. Bradley and O.L. Mangasarian. Massive data discrimination via linear support vector machines. *Optimization Methods and Software*, 13:1–10, 2000. DOI: 10.1080/10556780008805771 28

C. Burges. A tutorial on support vector machines for pattern recognition. *Data Mining and Knowledge Discovery*, 2:121–167, 1998. DOI: 10.1023/A:1009715923555 33

C. Campbell and K.P. Bennett. A linear programming approach to novelty detection. In *Advances in Neural Information Processing Systems, 14*, pages 395–401. MIT Press, 2001. 33

C. Campbell, N. Cristianini, and A. Smola. Query learning with large margin classifiers. In *Proceedings of ICML2000*, pages 111–118, 2000. 25

O. Chapelle and V. Vapnik. Model selection for support vector machines. In *Advances in Neural Information Processing Systems, 12*, pages 673–680. MIT Press, 2000. 50

O. Chapelle, V. Vapnik, O. Bousquet, and S. Mukerjhee. Choosing multiple parameters for SVM. *Machine Learning*, 46:131–159, 2002. DOI: 10.1023/A:1012450327387 62

C. Cortes and V. Vapnik. Support vector networks. *Machine Learning*, 20:273–297, 1995. DOI: 10.1023/A:1022627411411 14

N. Cristianini, C. Campbell, and J. Shawe-Taylor. Dynamically adapting kernels in support vector machines. In *Advances in Neural Information Processing Systems, 11*, pages 204–210. MIT Press, Cambridge, MA, 1999. 50

F. Cucker and D.X. Zhou. *Learning Theory: An Approximation Theory Viewpoint*. Cambridge University Press, 2007. 8

T. Damoulas and M. Girolami. Probabilistic multi-class multi-kernel learning: on protein fold recognition and remote homology detection. *Bioinformatics*, 24:1264–1270, 2008. DOI: 10.1093/bioinformatics/btn112 64

T.G. Dietterich and G. Bakiri. Solving multiclass learning problems via error-correcting output codes. *Journal of Artificial Intelligence*, 2:263–286, 1995. DOI: 10.1613/jair.105 8

K.-B. Duan and S.S. Keerthi. Which is the best multiclass SVM method? an empirical study. *Lecture Notes in Computer Science*, 3541:278–285, 2006. DOI: 10.1007/11494683_28 8

T. Gartner, P. Flach, and S. Wrobel. On graph kernels: Hardness results and efficient alternatives. In *Proceedings Annual Conference Computational Learning Theory (COLT)*, pages 129–143. Springer, 2003. 56

T. Graepel, R. Herbrich, P. Bollmann-Sdorra, and K. Obermayer. Classification on pairwise proximity data. In *Advances in Neural Information Processing Systems, 11*, pages 438–444. MIT Press, 1998. 47

T. Graepel, R. Herbrich, B. Schölkopf, A.J. Smola, P.L. Bartlett, K. Muller, K. Obermayer, and R.C. Williamson. Classification on proximity data with LP-machines. In *Ninth International Conference on Artificial Neural Networks*, volume 470, pages 304–309, 1999. DOI: 10.1049/cp:19991126 28

I. Guyon and A. Elisseeff. An introduction to variable and feature selection. *Journal of Machine Learning Research*, 3:1157–1182, 2003. http://www.jmlr.org 21

I. Guyon, J. Weston, S. Barnhill, and V. Vapnik. Gene selection for cancer classification using support vector machines. *Machine Learning*, 46:389–422, 2002. DOI: 10.1023/A:1012487302797 21

E. Harrington, R. Herbrich, J. Kivinen, J.C. Platt, , and R.C. Williamson. Online Bayes point machines. In *Proceedings of the Seventh Pacific-Asia Conference on Knowledge Discovery and Data Mining*. Springer-Verlag, 2003. DOI: 10.1007/3-540-36175-8_24 28

T. Hastie and R. Tibshirani. Classification by pairwise coupling. *The Annals of Statistics*, 26:451–471, 1998. DOI: 10.1214/aos/1028144844 8

R. Herbrich, T. Graepel, and C. Campbell. Bayes point machines. *Journal of Machine Learning Research*, 1:245–279, 2001. http://www.jmlr.org 28

R. Hettich and K.O. Kortanek. Semi-infinite programming: Theory, methods and applications. *SIAM Review*, 3:380–429, 1993. DOI: 10.1137/1035089 62, 67

A.E. Hoerl and R. Kennard. Ridge regression: biased estimation for nonorthogonal problems. *Technometrics*, 12:55–67, 1970. DOI: 10.2307/1267352 40

T. Joachims. Estimating the generalization performance of an SVM efficiently. In *Proceedings of the Seventeenth International Conference on Machine Learning*, pages 431–438. Morgan Kaufmann, 2000. 50

H. Kashima, K. Tsuda, and A. Inokuchi. Kernels on graphs. In *Kernels and Bioinformatics*, pages 155–170. MIT Press, Cambridge, MA, 2004. 56

S. Keerthi, S. Shevade, C. Bhattacharyya, and K. Murthy. Improvements to platt's SMO algorithm for SVM classifier design. *Neural Computation*, 13:637–649, 2001. DOI: 10.1162/089976601300014493 17

M. Kloft, U. Brefeld, S. Sonnenburg, P. Laskov, K.-R. Müller, and A. Zien. Efficient and accurate LP-norm multiple kernel learning. In *Advances in Neural Information Processing Systems, 22*, pages 997–1005. MIT Press, 2009. 64

I. R. Kondor and J. D. Lafferty. Diffusion kernels on graphs and other discrete structures. In *Proceedings of the International Conference on Machine Learning*, pages 315–322. Morgan Kaufmann, San Francisco, CA, 2002. 56

G. R. G. Lanckriet, N. Cristianini, P. Bartlett, L. El Ghaoui, and M.I. Jordan. Learning the kernel matrix with semidefinite programming. *Journal of Machine Learning Research*, 5:27–72, 2004a. 57, 58, 64

G. R. G. Lanckriet, T. De Bie, N. Cristianini, M.I. Jordan, and W.S. Noble. A statistical framework for genomic data fusion. *Bioinformatics*, 2004b. DOI: 10.1093/bioinformatics/bth294 64

Y. Lee, Y. Lin, and G. Wahba. Multicategory support vector machines, 2001. DOI: 10.1198/016214504000000098 8

C. Lemaréchal and C. Sagastizabal. Practical aspects of moreau-yosida regularization : theoretical preliminaries. *SIAM Journal of Optimization*, 7:867–895, 1997. DOI: 10.1137/S1052623494267127 62

C. Leslie and R. Kuang. Fast kernels for inexact string matching. *Lecture Notes in Computer Science*, 2777:114–128, 2003. DOI: 10.1007/978-3-540-45167-9_10 53

H. Lodhi, C. Saunders, J. Shawe-Taylor, N. Cristianini, and C. Watkins. Text classification using string kernels. *Journal of Machine Learning Research*, 2:419–444, 2002. http://www.jmlr.org 53

D.G. Luenberger. *Linear and Nonlinear Programming*. Addison Wesley, 2003. 67

R. Luss and A. d'Aspremont. Support vector machine classification with indefinite kernels. In *Advances in Neural Information Processing Systems, 20*, pages 953–960. MIT Press, 2008. 47

A.M. Malyscheff and T.B. Trafalis. An analytic center machine for regression. *Central European Journal of Operational Research*, 10:297–334, 2002. 28

O.L. Mangasarian. Linear and nonlinear separation of patterns by linear programming. *Operations Research*, 13:444–452, 1965. DOI: 10.1287/opre.13.3.444 28

C. A. Micchelli and M. Pontil. Learning the kernel function via regularization. *Journal of Machine Learning Research*, 9:1049–1125, 2005. 57

S. Mika, G. Ratsch, J. Weston, B. Schölkopf, and K.-R. Muller. Fisher discriminant analysis with kernels. In *Neural networks for signal processing IX*, pages 41–48, 1999. DOI: 10.1109/NNSP.1999.788121 28

T. Minka. *A family of algorithms for approximate Bayesian inference*. PhD thesis, MIT, 2001. 28

Y. Nesterov. *Introductory Lectures on Convex Optimization: A Basic Course*. Springer, 2003. 62

Y. Nesterov and A. Nemirovsky. *Interior Point Polynomial Methods in Convex Programming: Theory and Algorithms*. SIAM, 1994. 60

E. Pekalska, P. Paclik, and R.P.W. Duin. A generalized kernel approach to dissimilarity based classification. *Journal of Machine Learning Research*, 2:175–211, 2002. http://www.jmlr.org 47

J. Platt. Fast training of SVMs using sequential minimal optimization. In C. Burges B. Schölkopf and A. Smola, editors, *Advances in Kernel Methods: Support Vector Learning*. MIT press, Cambridge, MA, 1999a. 17

J. Platt. Probabilistic outputs for support vector machines and comparison to regularised likelihood methods. In *Advances in large margin classifiers*, pages 61–74. MIT Press, 1999b. 8, 29

J. Platt, N. Cristianini, and J. Shawe-Taylor. Large margin DAGS for multiclass classification. In *Advances in Neural Information Processing Systems, 12*, pages 547–553. MIT Press, Cambridge, MA, 2000. 8

A. Rakotomamonjy, F. Bach, S. Canu, and Y. Grandvalet. SimpleMKL. *Journal of Machine Learning Research*, 9:2491–2521, 2008. 57, 62

C.E. Rasmussen and C. Williams. *Gaussian Processes for Machine Learning*. The MIT Press, 2006. 30

V. Roth and V. Steinhage. Nonlinear discriminant analysis using kernel functions. In *Advances in Neural Information Processing Systems, 12*, pages 568–574. MIT Press, 2000. 28

V. Roth, J. Laub, M. Kawanabe, and J.M. Buhmann. Optimal cluster preserving embedding of nonmetric proximity data. *IEEE Transactions on Pattern Analysis and Machine Intelligence*, 25: 1540–1551, 2003. DOI: 10.1109/TPAMI.2003.1251147 47

P. Rujan and M. Marchand. Computing the Bayes kernel classifier. In *Advances in Large Margin Classifiers*, pages 329–347. MIT Press, 2000. 28

C. Saunders, A. Gammermann, and V. Vovk. Ridge regression learning algorithm in dual variables. In J. Shavlik, editor, *Proceedings of the Fifteenth International Conference in Machine Learning (ICML)*. Morgan Kaufmann, 1998. 40

B. Schölkopf and A. Smola. *Learning with Kernels*. The MIT Press, Cambridge, MA, 2002a. 17, 28, 29, 30, 33

B. Schölkopf and A.J. Smola. *Learning with Kernels*. The MIT Press, Cambridge, MA, USA, 2002b. 40

B. Schölkopf, C. Burges, and V. Vapnik. Extracting support data for a given task. In U.M. Fayyad and R. Uthurusamy, editors, *Proceedings: First International Conference on Knowledge Discovery and Data Mining*. AAAI Press, Menlo park, CA, 1995. 33

B. Schölkopf, J.C. Platt, J. Shawe-Taylor, A.J. Smola, and R.C. Williamson. Estimating the support of a high-dimensional distribution. In *Microsoft Research Corporation Technical Report MSR-TR-99-87*, 1999. DOI: 10.1162/089976601750264965 33

B. Schölkopf, A. Smola, R.C. Williamson, and P.L. Bartlett. New support vector algorithms. *Neural Computation*, 12:1207–1245, 2000. DOI: 10.1162/089976600300015565 14

P. Schölkopf, B. and Bartlett, A. Smola, and R. Williamson. Support vector regression with automatic accuracy control. In M. Boden L. Niklasson and T. Ziemke, editors, *Proceedings of the 8th International Conference on Artificial Neural Networks, Perspectives in Neural Computing*. Springer Verlag, Berlin, 1998. 40

J. Shawe-Taylor and N. Cristianini. *Kernel Methods for Pattern Analysis*. Cambridge University Press, 2004. 29, 53

S. K. Shevade, S. S. Keerthi, C. Bhattacharyya, and K. R. Krishna Murthy. Improvements to the SMO algorithm for SVM regression. *IEEE Transactions on Neural Networks*, 11:1188–1194, 2000. DOI: 10.1109/72.870050 17

A.J. Smola and I.R. Kondor. Kernels and regularization on graphs. In *Lecture Notes in Computer Science*, pages 144–158. Springer-Verlag, Heidelberg, Germany, 2003. DOI: 10.1007/978-3-540-45167-9_12 56

A.J. Smola and B. Schölkopf. A tutorial on support vector regression. *Statistics and Computing*, 14: 199–222, 2004. DOI: 10.1023/B:STCO.0000035301.49549.88 40

P. Sollich. Bayesian methods for support vector machines: Evidence and predictive class probabilities. *Machine Learning*, 46:21–52, 2002. DOI: 10.1023/A:1012489924661 50

S. Sonnenburg, G. Rätsch, C. Schäfer, and B. Schölkopf. Large scale multiple kernel learning. *Journal of Machine Learning Research*, 7:1531–1565, 2006. 57, 62

S. Sonnenburg, G Ratsch, and K. Rieck. *Large-Scale Learning with String Kernels*, pages 73–103. MIT Press, 2007. 54

I. Steinwart and A. Christmann. *Support Vector Machines*. Springer, 2008. 8

J.F. Sturm. Using SEDUMI 1.02a, a matlab toolbox for optimization over symmetric cones. *Optimization Methods and Software*, pages 625–653, 1999. DOI: 10.1080/10556789908805766 60

J.A.K. Suykens and J. Vandewalle. Least squares support vector machine classifiers. *Neural Processing Letters*, 9:293–300, 1999. DOI: 10.1023/A:1018628609742 28

B. Taskar, V. Chatalbashev, D. Koller, and C. Guestrin. Learning structured prediction models: a large margin approach. In *Proceedings of the 22nd International conference on machine learning*, 2005. DOI: 10.1145/1102351.1102464 43

D. Tax and R. Duin. Data domain description by support vectors. In M. Verleysen, editor, *Proceedings of ESANN99*, pages 251–256. D. Facto Press, Brussels, 1999. 33

D. Tax, A. Ypma, and R. Duin. Support vector data description applied to machine vibration analysis. In M. Boasson, J. Kaandorp, J. Tonino, and M. Vosselman, editors, *Proc. 5th Annual Conference of the Advanced School for Computing and Imaging*, pages 398–405. Heijen, NL, 1999. 33

S. Tong and D. Koller. Support vector machine active learning with applications to text classification. *Journal of Machine Learning Research*, 2:45–66, 2001. http://www.jmlr.org 25

T.B. Trafalis and A.M. Malyscheff. An analytic center machine. *Machine Learning*, 46:203 – 224, 2002. DOI: 10.1023/A:1012458531022 28

I. Tsochantaridis, T. Joachims, T. Hofmann, and Y. Altun. Large margin methods for structured and independent output variables. *Journal of machine learning research*, 6:1453–1584, 2005. 43

UCI Machine learning Repository. http://www.ics.uci.edu/~mlearn/mlrepository.html. 10

L. Vandenberghe and S. Boyd. Semidefinite programming. *SIAM Review*, 38, 1996. DOI: 10.1137/1038003 60, 62, 67

V. Vapnik. *The Nature of Statistical Learning Theory*. Springer, New York, 1995. 40

V. Vapnik. *Statistical Learning Theory*. Wiley, 1998. 7, 40

V. Vapnik and O. Chapelle. Bounds on error expectation for support vector machines. *Neural Computation*, 12:2013–2036, 2000. DOI: 10.1162/089976600300015042 50

K. Veropoulos, C. Campbell, and N. Cristianini. Controlling the sensitivity of support vector machines. In *Proceedings of the International Joint Conference on Artificial Intelligence (IJCAI)*, 1999. 14

S. Vishwanathan and A Smola. Fast kernels for string and tree matching. In *Advances in Neural Information Processing Systems, 15*, pages 569–576. MIT Press, 2003. 53

S. V. N. Vishwanathan, K. M. Borgwardt, I. R. Kondor, and N. N. Schraudolph. Graph kernels. *Journal of Machine Learning Research*, 9:1–41, 2008. 56

M.K. Warmuth, J. Liao, G. Ratsch, M. Mathieson, S. Putta, and C. Lemmen. Active learning with support vector machine in the drug discovery process. *J. Chem. Inf. Comput. Sci.*, 43:667–673, 2003. DOI: 10.1021/ci025620t 25

J. Weston and C. Watkins. Multi-class support vector machines. In M Verleysen, editor, *Proceedings of ESANN99*, pages 219–224. D. Facto Press, Brussels, 1999. 8

J. Weston, A. Gammerman, M. Stitson, V. Vapnik, V. Vovk, and C. Watkins. Support vector density estimation. In C. Burges B. Schölkopf and A. Smola, editors, *Advances in Kernel Methods: Support Vector Machines*, pages 293–306. MIT Press, Cambridge, MA, 1998. 40

R.D. Williams, S.N. Hing, B. T. Greer, Craig C. Whiteford, J. S. Wei, R. Natrajan, A. Kelsey, S. Rogers, C. Campbell, K. Pritchard-Jones, and J. Khan. Prognostic classification of relapsing favourable histology wilms tumour using cdna microarray expression profiling and support vector machines. *Genes, Chromosomes and Cancer*, 41:65 – 79, 2004. DOI: 10.1002/gcc.20060 21

Z. Xu, R. Jin, I. King, and M.R. Lyu. An extended level method for multiple kernel learning. In *Advances in Neural Information Processing Systems, 22*. MIT Press, 2008. 62

Y. Ying and C. Campbell. Generalization bounds for learning the kernel. In *Proceedings of 22nd Annual Conference on Learning Theory (COLT)*, 2009. 57

Y. Ying and D.X. Zhou. Learnability of gaussians with flexible variances. *Journal of Machine Learning Research*, 8:249–276, 2007. 57

Y. Ying, C. Campbell, and M. Girolami. Analysis of SVM with indefinite kernels. In *Advances in Neural Information Processing Systems, 22*, pages 2205–2213. MIT Press, 2009a. 47, 62

Y. Ying, K. Huang, and C. Campbell. Enhanced protein fold recognition through a novel data integration approach. *BMC Bioinformatics*, 10:267–285, 2009b. DOI: 10.1186/1471-2105-10-267 62, 64

A.L. Yuille and A. Rangarajan. The concave convex procedure. *Neural Computation*, 15:915–936, 2003. DOI: 10.1162/08997660360581958 43

Authors' Biography

COLIN CAMPBELL

Dr. Colin Campbell holds a BSc degree in physics from Imperial College, London, and a PhD in mathematics from King's College, London. He joined the Faculty of Engineering at the University of Bristol in 1990 where he is currently a Reader. His main interests are in machine learning and algorithm design. Current topics of interest include kernel-based methods, probabilistic graphical models and the application of machine learning techniques to medical decision support and bioinformatics. His research is supported by the EPSRC, Cancer Research UK, the MRC and PASCAL2.

YIMING YING

Dr. Yiming Ying received his BSc degree in mathematics from Zhejiang University (formally, Hangzhou University) in 1997 and his PhD degree in mathematics from Zhejiang University in 2002, Hangzhou, China. He is currently a Lecturer (Assistant Professor) in Computer Science in the College of Engineering, Mathematics and Physical Sciences at the University of Exeter, UK. His research interests include machine learning, pattern analysis, convex optimization, probabilistic graphical models and applications to bioinformatics and computer vision.